AF380499

Surface Wetting

Kock-Yee Law • Hong Zhao

Surface Wetting

Characterization, Contact Angle, and Fundamentals

 Springer

Kock-Yee Law
Founder & CEO at Research
  and Innovative Solutions
Penfield, NY, USA

Hong Zhao
School of Engineeing, Mechanical and
  Nuclear Engineering
Virginia Commonwealth University
Richmond, VA, USA

ISBN 978-3-319-25212-4   ISBN 978-3-319-25214-8   (eBook)
DOI 10.1007/978-3-319-25214-8

Library of Congress Control Number: 2015953774

Springer Cham Heidelberg New York Dordrecht London
© Springer International Publishing Switzerland 2016

Printed on acid-free paper

Springer International Publishing AG Switzerland is part of Springer Science+Business Media
(www.springer.com)

Preface

Surface science involves studies of properties and characteristics of solid surfaces and their interactions with liquids. It is a multidiscipline field with great academic interests and tremendous applications in the industry. Unfortunately, the field of surface has been full of controversies and misconceptions. The Young's equation is widely credited as an initiator of surface research. However, it is also a source for continuous arguments and debates. I and Hong Zhao stumbled into this field seriously about 8 years ago with the objectives of understanding how inks adhere on and release from different printing surfaces during the printing process. It was our hope that fundamental understandings of various ink–surface interactions would not only lead to improvement in the printing process but also enable better ink and surface material designs. However, we found the literature messy and not very helpful. Our literature study yielded limited guidance and the messages obtained at times are confusing. There was no standardized measurement protocol. We often found that definitions were unclear and terminologies were created arbitrarily. In our lab, the structure–property relationship only followed about 50 % of the time and unexpected observation was always classified as exception. This certainly does not sound like a mature science field. We were actually not alone because many academic colleagues also echo these shortfalls. This has been the main motivation and passion for us to write this book. We would like to share some fundamental basic concepts we have learned through this journey. This book is not intended for expert researchers who may view the content as nothing new. Rather, it is intended for senior undergraduate and graduate students in various science and engineering fields as well as researchers like us who have the need to get into the field of surface in their jobs.

Penfield, NY
Richmond, VA, USA

Kock-Yee Law
Hong Zhao

Contents

Chapter 1
Background

Abstract Studies of surface and liquid–solid interaction have always been an important branch of science, and its role just increases exponentially due to the expanded application of digital printing. To date, due to the on-demand nature of ink printing, it has become a manufacturing technology for many current and futuristic electronic devices, such as display, printed electronics, and wearable and flexible devices. Research on surface has always been messy, however. Debates and rigorous discussion on the Young's contact angle, measurement procedures, and data interpretation have been ongoing in the surface literature for many decades. In this chapter, the justification of writing this book is described. The shortfalls in surface science are briefly overviewed. A roadmap that systematically addresses fundamental issues on measurements, basic concepts in wetting and surface characterization, and definitions and terminologies is provided throughout this book. It is our hope that this collection of surface fundamentals will improve readers' basic understanding of all the key concepts, which will eventually enhance the quality of surface research in the future.

Keywords Surface • Young's equation • Contact angle • Wetting • Liquid–solid interaction • Young's angle • Contact angle measurement • Data interpretation

Surface is a very important branch of science that touches all facets of our lives. Fundamental understanding of the interactions between a liquid droplet and a solid surface, such as wetting, spreading, adhesion, and de-wetting, is not only crucial to the science itself, but also of tremendous values to many seemingly unrelated applications. In our daily activities such as cleaning and washing, surface active materials known as detergents are commonly used to aid detachments of dirt and stain from a solid surface initiating the cleaning process. Although we have been taken the process for granted, there is a lot of science involved in washing and cleaning, e.g., the type of detergent used, its efficiency, cost, and the impact to human and environment. Even as simple as ink writing with a fountain pen, Kim and co-workers [1] found that the line width of the ink image on paper depends on the speed of the pen as well as the physicochemical properties of both ink and paper. Controlling ink wetting and spreading is crucial in avoiding clogging and ink spreading. Similarly, mastering and controlling liquid–solid interactions have become critical skills in many industries, such as in the design and manufacturing of paints, stain/

© Springer International Publishing Switzerland 2016
K.-Y. Law, H. Zhao, *Surface Wetting*, DOI 10.1007/978-3-319-25214-8_1

Fig. 1.1 A schematic of the offset printing process

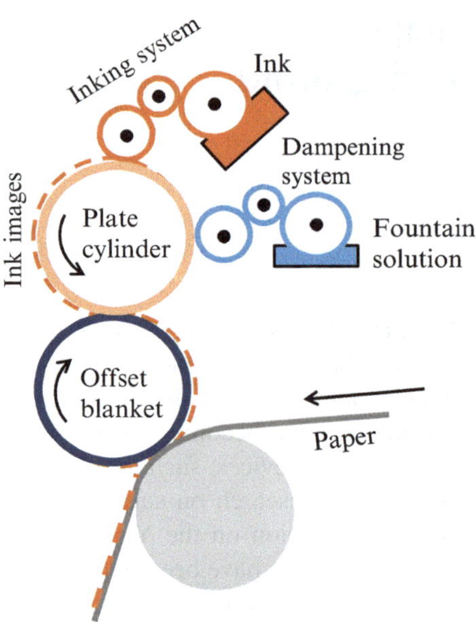

soil-resistant textiles and clothing, anticorrosion surfaces for bridges and other metal structures, antifouling coatings for ships and marine structures, anti-icing surfaces for power line, airplane and roof-top, mining, and petroleum fracking. Whether it is surface design or process development, knowledge in manipulating liquid–surface interaction in the "right" place at the "right" time is imperative in microfluidic device [2, 3], oil/fluid transportation [4], and printing and coating [5, 6], just to name a few. In printing, understanding the wetting and de-wetting of ink on different printing surfaces is critical to the quality of the final print. As a society, we have been practicing offset printing for more than a century, and the entire print process is a good illustration on the importance of ink–surface interaction [7]. A schematic for the offset printing process is shown in Fig. 1.1.

In a very simple term, the offset printing process involves (1) wetting of the plate cylinder with a fountain solution image wise through the dampening system, followed by (2) inking the plate cylinder with the inking system, (3) transfer of the ink image from the plate cylinder to an offset blanket, and (4) fixing it on paper. In the first step, the desirable outcome is to have the fountain solution first wets and then pins on the surface of the plate cylinder. While wetting is required for the formation of the wetted images, a successful pinning of the contact line on the plate cylinder is needed to ensure image integrity and resolution control. Offset inks are acrylate materials, and they will be repelled by areas that are wetted with the fountain solution. On the other hand, ink will be attracted and adhered to the oleophilic areas on the plate cylinder surface. Controlling ink adhesion in the oleophilic region and repelling it from the fountain solution wetted areas are critical to the image quality of the output. Afterward, the transfer of the ink image to the offset blanket and then paper is more straightforward as these surfaces are selected based on their relative affinity to the ink materials. Although printing (on papers) is an industry in decline, printing has evolved and has become a manufacturing technology for printed electronics, flexible and wearable devices, ceramics, textiles, solar cells, and many others [8–20]. Arias et al. [19] showed that balance between pinning and overspreading of printed liquid ink on the solid surface is very important in defining the position, resolution, and size in the fabrication of thin-film-transistor array. Jetted ink drop is also known to form the so-called coffee ring stain due to un-optimized spreading and drying, which is detrimental to the quality of the printed image and ultimately the performance of the printed device [21]. The need of characterizing surfaces and understanding liquid–surface interaction continues to play a critical role in the modern technology arena [8].

Studies of surface and liquid–surface interaction can be traced back to Thomas Young's work two centuries ago [22]. In his legendary article entitled "An Essay on the Cohesion of Fluids," he described his study of wetting of glass by water and mercury, wetting of various metal surfaces by mercury and the capillary effect. He descriptively stated that the angle of contact between a liquid on a solid surface is the result of a mechanical equilibrium among the three surface tensions at the contact line. They are the liquid, solid and liquid–solid interfacial surface tensions. This has become the famous Young's equation in the literature. Today, surface is a multidiscipline subject and is studied by chemists, physicists, and engineers both theoretically and experimentally worldwide. Research topics range from fundamental understanding of the wetting, spreading, adhesion, and de-wetting phenomena to their applications in materials, surface coatings, and device innovations. Unfortunately, there have been continuous confusion and faulty intuition about the measurements, data interpretation, and definitions in the surface literature. One of the reasons is due to the multidiscipline nature in this field, where researchers with a very diverse background are investigating surfaces together. Another reason, which was pointed out by Gao and McCarthy, is their insufficient basic surface training in school [23, 24]. This flaw in surface science has been recognized by expert researchers. In 2009, Gao and McCarthy published an excellent article titled "Wetting 101°" where they discussed the faulty perception and used demonstrative examples to illustrate the correct basic concepts [24]. While we are certainly benefitted from the article and references therein, the faulty perception and confusion are continuing unfortunately.

Contact angle measurement is commonly used to characterize a surface and to study various wetting and de-wetting phenomena. While the measurement is simple, the interpretation is not. This point has been noted by many surface investigators in the past, e.g., Pease in 1945 [25], Morra et al. [26] in 1990, Kwok and Neumann [27] in 1999 and more recently by Marmur [28] as well as Strobel and Lyons [29]. Prior to data interpretation, one has to make sure that the measurement apparatus and procedures are impeccable. Over the years, many have voiced concerns over surface preparation and conditioning, measurement procedure and technique, and data analysis [26, 30–35]. It is therefore imperative for the community to have a set of common measurement protocol or guideline, so that inter laboratory data can be compared. Discrepancy in conclusion can be rationalized without concerns of experimental setups or procedures.

In view of the ongoing discussions on measurement procedures, data interpretation, terminologies, and definitions, the surface community would be benefitted for an overview of the past conversations and a recent account of the resolution. The objective of this book is to provide a coherent, easy to understand, fundamental picture on surface wetting and characterization. In Chap. 2, we first summarize the best practices in static and dynamic contact angle measurements. This may be served as a standard protocol for surface researchers and practitioners in the future. From there, data collected in different laboratories can be compared without casting doubt to each other. Some of the fundamental measurement issues, such as drop size, drop dispensing procedure, lab ambient condition (temperature and relative humidity), and method of calculating the contact angle will be discussed. This is followed by discussions of the concept of wetting, first on smooth surfaces in Chap. 3 and then on rough surfaces in Chap. 4. Important concepts such as (1) the Young's angle is a result of a mechanical equilibrium, not thermodynamic equilibrium, at the three phase contact line, (2) the liquid droplets are all in their metastable states in their static, advancing and receding positions, and (3) contact line, not contact area determines the contact angle, will be articulated with conclusions that are supported by both experimental and theoretical data. The rationale for the shortfall in both the Wenzel and Cassie–Baxter analysis of wetting on rough and porous surfaces is summarized. Recent results on factors that govern wettability and wetting dynamics of liquid on rough surfaces are overviewed. Visuals on the distortion of the contact line by surface roughness as well as the structure of the liquid–solid–air composite interfaces are reported. Chapter 4 also includes discussions on the design principles for superhydrophobic and superoleophobic surfaces as well as the challenges related to technology implementation.

Due to simplicity of the contact angle measurement, it has become a popular tool to characterize the property of a surface or probing liquid–solid interactions. However, the literature is full of conflicting information owing to the difficulty in correctly interpreting contact angle data. Chapter 5 is devoted to provide experimental data to shrine light into this specific issue. Evidence is provided that advancing and receding contact angles are measurement of surface wettability and adhesion, respectively. The stickiness of liquid on surface can be predicted from the sliding angle. A recommendation for surface characterization is provided. As for contact angle hysteresis, the difference between advancing and receding contact angle, it is shown to relate to adhesion and stickiness qualitatively only. More work is needed to clarify the origin of contact angle hysteresis as well as its role in surface characterization. As a fundamental book, it is hard not to discuss the "pains" we have in surface definitions and terminologies as well as early work on the use of contact angle to determine the surface energy of solid, where the practice and its usefulness have been constantly challenged. In Chaps. 6 and 7, updates on the definitions for hydrophilicity/hydrophobicity, oleophilicity/oleophobicity, and other related terminologies will be provided. The different methodologies used to determine surface energy will briefly be reviewed. Fundamental issues will be discussed, and a path to move forward is shared. We are not taking side in these discussions, rather updating the readers with the latest experimental data and the current status.

This book will conclude with a brief look back on the evolution of surface science, followed by a summary of some of the basic concepts as a reminder for new or young researchers in this field. It is our hope that surface science will prosper when researchers in the next-generation are armed with improved basic concepts and fundamentals.

References

1. Kim J, Moon MW, Lee KR, Mahadevan L, Kim HY (2011) Hydrodynamics of writing with ink. Phys Rev Lett 107:264501
2. Bruzewicz DA, Reches M, Whitesides GM (2008) Low-cost printing of PDMS barriers to define microchannels in paper. Anal Chem 80:3387–3392
3. Martinez AW, Phillips ST, Whitesides GM (2008) Three-dimensional microfluidic devices fabricated in layered paper and tape. Proc Natl Acad Sci U S A 105:19606–19611
4. Paso K, Kompalla T, Aske N, Ronningsen HD, Oye G, Sjoblom J (2009) Novel surfaces with applicability for preventing wax deposition: a review. J Dispersion Sci Technol 30:757–781
5. Katano Y, Tomono H, Nakajima T (1994) Surface property of polymer films with fluoroalkyl side chains. Macromolecules 27:2342–2344
6. Samuel B, Zhao H, Law KY (2011) Study of wetting and adhesion interactions between water and various polymer and superhydrophobic surfaces. J Phys Chem C 115:14852–14861
7. Dejidas LP Jr, Destree TM (2006) Sheetfed offset press operating, 3rd edn. PIA-GATF Press, Pittsburgh, Chapter 6
8. Perelaer J, Smith PJ, Mager D, Soltman D, Volkman SK, Subramanian V, Korvink JG, Schubert US (2010) Printed electronics: the challenges involved in printing devices, interconnects, and contacts based on inorganic materials. J Mater Chem 20:8446–8453
9. Zielke D, Hubler AC, Hahn U, Brandt N, Bartzsch M (2005) Polymer-based organic field-effect transistor using off-set printed source-drain structures. Appl Phys Lett 87:123508
10. Kang B, Lee WH, Cho K (2013) Recent advances in organic transistor printing processes. ACS Appl Mater Interfaces 5:2302–2315
11. Fukuda K, Sekine T, Kumake D, Tokito S (2013) Profile control of inkjet printed silver electrodes and their application to organic transistors. ACS Appl Mater Interfaces 5:3916–3920
12. Schuppert A, Thielen M, Reinhold J, Schmidt WA, Schoeller F, Osnabruck O (2011) Ink jet printing of conductive silver tracks from nanoparticle inks on mesoporous substrates. NIP27 and digital fabrication 2011. Technical Programs and Proceeding. pp 437–440
13. Wu Y (2011) Inkjet printed silver electrodes for organic thin film transistors. NIP27 and digital fabrication 2011. Technical Programs and Proceeding. pp 441–444
14. Zipperer D (2011) Printed electronic for flexible applications. NIP27 and digital fabrication 2011. Technical Programs and Proceeding. pp 452–453
15. Simske S, Aronoff JS, Duncan B (2011) Printed antennas for combined RFID and 2D barcodes. NIP27 and digital fabrication 2011. Technical Programs and Proceeding. pp 544–547
16. Sanz V, Bautista RY, Feliu C, Roig Y (2011) Technical evolution of ceramic tile digital decoration. NIP27 and digital fabrication 2011. Technical Programs and Proceeding. pp 532–536
17. Nossent KJ (2011) A breakthrough high speed wide format print concept for textile. NIP27 and digital fabrication 2011. Technical Programs and Proceeding. p 556
18. Huson D (2011) 3D printing of ceramics for design concept molding. NIP27 and digital fabrication 2011. Technical Programs and Proceeding. pp 815–818
19. Arias AC, Mackenzie JD, McCulloch I, Pivnay J, Salleo A (2010) Materials and applications for large area electronics: solution-based approaches. Chem Rev 110:3–24
20. Habas SE, Platt HAS, van Hest MFAM, Ginley DS (2010) Low-cost inorganic solar cells: from ink to printed devices. Chem Rev 110:6571–6594

21. Deegan RD, Bakajin O, Dupont TF, Huber G, Nagel SR, Witter TA (1997) Capillary flow as the cause of ring stains from dried liquid drops. Nature 389:827–829
22. Young T (1805) An essay on the cohesion of fluids. Phil Trans R Soc London 95:65–87
23. Gao LC, McCarthy TJ (2009) An attempt to correct the faulty intuition perpetuated by the Wenzel and Cassie "Laws". Langmuir 25:7249–7255
24. Gao LC, McCarthy TJ (2009) Wetting 101°. Langmuir 25:14105–14115
25. Pease DC (1945) The significance of the contact angle in relation to the solid surface. J Phys Chem 49:107–110
26. Morra M, Occhiello E, Garbossi F (1990) Knowledge about polymer surfaces from contact angle measurements. Adv Colloid Interface Sci 32:79–116
27. Kwok DY, Neumann AW (1999) Contact angle measurement and contact angle interpretation. Adv Colloid Interfacial Sci 81:167–249
28. Marmur A (2006) Soft contact. Measurement and interpretation of contact angles. Soft Matter 2:12–17
29. Strobel M, Lyons SL (2011) An essay on contact angle measurements. Plasma Process Polym 8:8–13
30. Bartell FE, Wooley AD (1933) Solid–liquid–air contact angles and their dependence upon the surface condition of the solid. J Am Chem Soc 55:3518–3527
31. Bartell FE, Hatch GB (1934) Wetting characteristics of galena. J Phys Chem 39:11–24
32. Fox HW, Zisman WA (1950) The spreading of liquids on low energy surfaces. 1. Polytetrafluoroethylene. J Colloid Sci 5:514–531
33. Good RJ (1977) Free energy of solids and liquids: thermodynamics, molecular forces, and structure. J Colloid Interface Sci 59:398–419
34. Decker EL, Frank B, Suo Y, Garoff S (1999) Physics of contact angle measurement. Colloids Surf A 156:177–189
35. Drelich J (2013) Guidelines to measurements of predictable contact angles using a sessile-drop technique. Surf Innov 1:248–254

Chapter 2
Contact Angle Measurements and Surface Characterization Techniques

Abstract Contact angle measurement has been an indispensable tool for surface characterization and wetting study due to its simplicity and versatility. In this chapter, major measurement techniques for static contact angle, sliding angle, and advancing/receding angle are overviewed. Critical procedural details including sessile-drop dispensing, drop size, drop profile capturing, and analysis are highlighted and discussed. Best practices on measuring sliding and advancing/receding angles are recommended with support of publications from leading researchers. It is our hope that the techniques described within will be used as guidelines for the research community.

Keywords Contact angle measurement · Goniometer · Static contact angle · Advancing contact angle · Receding contact angle · Sliding angle · Contact angle hysteresis · Wilhelmy plate technique · Captive bubble method · Tilting plate method

2.1 Introduction

The notion of wettability was first described by Young two centuries ago [1]. Wetting of a solid surface by liquid can be quantitatively described from the profile of a liquid droplet, more specifically from the tangential angle at liquid–solid–air interface. This angle of contact θ is defined as the Young's angle or static contact angle (Fig. 2.1). It is a result of a mechanical equilibrium among the three surface tensions, the liquid surface tension (γ_{LV}), the solid surface tension (γ_{SV}), and the liquid–solid interfacial tension (γ_{SL}) and is expressed as the Young's equation to date.

$$\gamma_{SV} = \gamma_{LV} \cdot \cos \theta + \gamma_{SL} \tag{2.1}$$

According to Pease [2] and later Morra et al. [3], Rayleigh [4] and Bartell [5] had observed two contact angles besides the Young's angle that could be measured reliably with liquid advancing and receding at a very slow rate. These are the advancing and receding angles we know to date, and their difference is hysteresis and has become a signature of real surface during that research era. As procedure and instrument are improved, measuring the static contact angle by the sessile drop method

© Springer International Publishing Switzerland 2016
K.-Y. Law, H. Zhao, *Surface Wetting*, DOI 10.1007/978-3-319-25214-8_2

Fig. 2.1 Graphic vector representation of parameters in a sessile drop

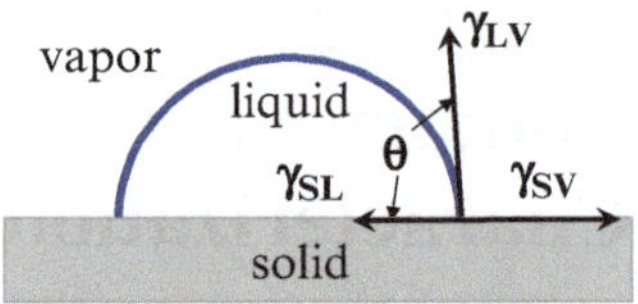

becomes more accurate and reproducible. Owing to simplicity of the technique, contact angle measurement has become a very popular tool for surface characterization and studies of various liquid–solid interactions. As mentioned in Chap. 1, the interpretation of contact angle has not been straightforward and has been a subject of constant debate in the literature. There are multiple contributors to this debate, and they include methods and conditions of generating consistent reliable data, comprehension of the basic wetting concepts, and sometimes faulty intuitions. In this chapter, the apparatus and the procedures for various contact angle measurements are reviewed with special attention to the critical details that were of concerns in the past. Most of these critical details turn out to be a source of data variability. A thorough understanding of these details coupled with the best work practice would allow the community to produce trustworthy reproducible data. When that stage is reached, inter-laboratory data can be compared with confidence and data interpretation and conclusion can be drawn without concerns of doubtful inputs. At the end of this chapter, a set of general procedures useful for routine surface characterization and wetting study is summarized. Basic concepts on the wetting of smooth and rough surfaces will be presented and discussed in Chaps. 3 and 4. In Chap. 5, the focus is on connecting the contact angle data to wetting interactions and surface properties.

It is important to point out that most interests in surface are the learning of the chemical or physical properties of the solid surfaces and their interactions with liquids. This can simply be done by studying the static and dynamic contact angles of the liquid on the solid surface. *A key basic criterion for the study is that the liquid and the solid surface have to be nonreactive both physically and chemically.* The solid surface should be cleaned appropriately, and the liquid should be purified and free of surfactant-like contamination. Measurements should not be carried out if distortion of the surface dimension or liquid adsorption or dissolution of surface materials occurs. Liquid droplets are usually in the millimeter range. Since the angle of contact is captured optically, the solid surface has to be flat optically. The constituents on the solid surface have to be homogeneous relative to the size of the liquid droplet. Any microscopic heterogeneity, whether it is chemical or physical in origin, should be small and cause no distortion to the optical profile of the sessile droplet. Theoretical calculation on textured surface suggested that microscopic heterogeneity three orders of magnitudes smaller than the sessile droplet should have little effect on the optical profile [6]. In most practice, the diameter of the sessile drop is around 1–5 mm, suggesting that useful contact angle data can still be obtained with samples comprising microscopic heterogeneity on the order of 100–200 µm. We highly recommend using complemental surface characterization techniques, such as atomic force microscopy (AFM) or scanning electron microscopy (SEM), when studying microscopic heterogeneous flat or rough surfaces. Although small area samples and curved or super-rough surfaces are not covered in this chapter, we believe that these measurements still can be done by proper modification of the apparatus and procedure based on known principle.

Another important point is that contact angle measures macroscopic properties of the solid surface. Interactions between liquid and surface are molecular level events. The angle of contact on heterogeneous surface is still dictated by the chemistry of the local area at the microscopic level. The observed contact angle (some literature called this as apparent contact angle) is simply the sum of all contributing components, providing the heterogeneity is homogeneous with length scale ~3 orders of magnitude smaller than the probing liquid droplet.

The most widely adopted technique for measuring contact angles is the optical-based sessile drop method. A typical goniometer consists of a horizontal stage with a solid sample mount, located between a light source and a CCD camera. A motorized liquid dosing system dispenses a certain amount of testing liquid onto the solid surface, forming a sessile drop. The first contact angle goniometer was designed by William Zisman in 1960 and manufactured by ramé-hart [7]. The original manual contact angle goniometer used an eyepiece with a microscope and the contact angle was determined by a protractor (Fig. 2.2a). Since then, many goniometers have been designed by various manufactures with slightly different designs, accessories, and subsystems. The current generation of contact angle instruments uses cameras and software to capture and analyze the drop shape (Fig. 2.2b). In addition, today's goniometers usually have modular design, which enables the apparatus to be upgraded to accommodate additional capabilities, for example, high temperature environmental chamber, pressure/vacuum chamber, nano/pico dosing units, titling base, and full automation for drop dispensing for advancing and receding contact angle studies. If a high-speed camera system is installed, the dynamic of the wetting process can also be studied.

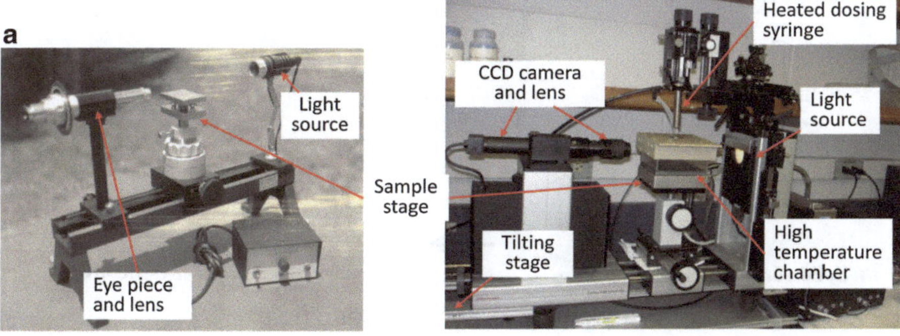

Fig. 2.2 Photographs of goniometers (**a**) classic ramé-hart model A100 [7] and (**b**) model OCA20 from Dataphysics

2.2 Determination of Static Contact Angle

2.2.1 Basic Measurement and Techniques

In most static contact angle measurement, the sessile drop is usually formed in open lab atmosphere by dispensing the testing liquid through a microsyringe gently onto a horizontal solid surface using a motor-controlled dosing unit. A schematic of the procedure is given in Fig. 2.3. As a general practice, both the test surface and the microsyringe should be appropriately cleaned. Particularly, the solid surface should be free of dirt, dust, and any contaminant. The test liquid should be best solvent grade, e.g., free of surfactant. Preferably, the entire apparatus should be on a vibration-free table, so that the drop formation is not affected by any vibrational noise from the surrounding environment. After the liquid droplet first contacts (wets) the surface, it spreads. As a rule of thumb, one should allow the drop to stabilize to reach its final static state before taking any measurement. For common liquids, such as water, hexadecane, and diiodomethane, this should take less than a second. For liquid with higher viscosity or new liquid/solid system, one can monitor the dynamic of the wetting process using the video system in the goniometer. When the liquid drop first contacts the surface, the dynamic contact angle is close to 180°, and the dynamic angle should decay to the steady state. From there, the drop profile is captured and analyzed. The static contact angle is obtained by curve-fitting the drop profile using vendor provided software. Details regarding drop shape analysis will be given in Sect. 2.2.4. The repeatability and accuracy of the measurement, to a large extent, depends on the consistency of the drop dispensing process, baseline determination on the sessile drop profile, and accuracy of the drop shape analysis.

The measurement and analysis for each liquid–solid system should be repeated 5–10 times in different area of the sample. For a well-behaved system, the variation among 5–10 measurements should be <2°. This will signal not only the procedures are reproducible, but the property of the surface is homogeneous and uniform.

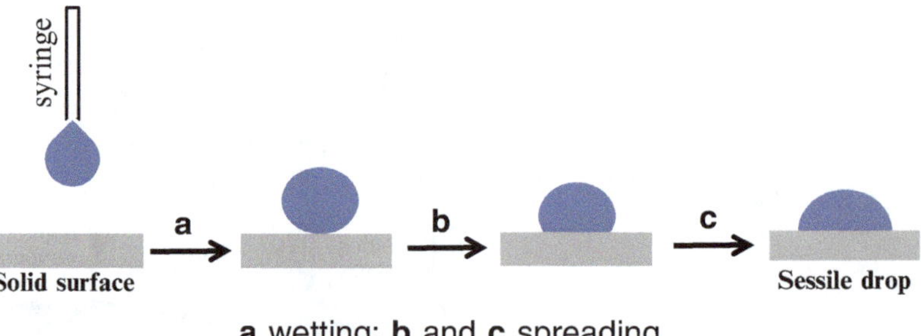

Fig. 2.3 A schematic showing the formation of a sessile droplet during static contact angle measurement

2.2.2 Best Practices and Critical Details

Throughout the development of surface science, researchers have not been thrilled seeing unreliable or inconsistent contact angle data [5, 8–11]. Good even recommended not to publish the paper if only one contact angle was reported, and he actually called for standardization of the measurement procedure ~40 years ago [10]. The issue involved is beyond cleanliness of the surface or purity of the solvent. Researchers are worried about the critical details during the course of the contact angle measurement, which may lead to variability and discrepancy between laboratories. Shuttleworth and Bailey [12] noted in 1948 that the final static position reaches by the sessile drop depends not only on the surface tensions of the liquid, solid, and liquid–solid interface, but also on the manner the drop is dispensed and the roughness of the surface. As our understanding of the wetting fundamentals is improved, we come to realize that the static state described by Young is actually a metastable wetting state (Sect. 3.3). The sensitivity of the measured static contact angle to experimental details is understandable. In this section, some of the critical details, which may not only affect the quality and consistency of the data, it can sometimes skew the trend being studied and leads to a different conclusion.

As mentioned above, a lot of measurement is carried out in open lab atmosphere. Even when the lab is conditioned (with controlled temperature and relative humidity), solvent still can evaporate. This is especially problematic with volatile liquids. When solvent is evaporated, the contact line starts to de-pin and recede. The contact angle is then technically not static. The drop is in the receding mode, but not necessarily in any equilibrium condition. This may introduce variability. On the other hand, we [13] as well as Zisman [9] found that contact angle measurements with water and hexadecane could be measured reliably and reproducibly (<2° error) in open lab environment. Zisman [9] did notice the solvent evaporation effect from low boiling point alkanes, such as pentane and hexane. If the use of volatile liquid is unavoidable, a simple way to eliminate data variability induced by solvent evaporation is to place the sample platform in a close chamber. By doing that, the liquid phase is in equilibrium with air and the temperature of the sessile drop and the vapor pressure of the solvent are constant throughout the measurement [11]. Better yet, if an environmental controlled unit such as that shown in Fig. 2.2b is used, one can not only maintain a constant temperature and vapor pressure, but also have the ability to regulate the measuring temperature anywhere from lab ambient (20 °C) to 400 °C.

Another frequently asked question is how long will it take for the sessile drop to reach the static state? As shall be discussed in Chap. 3, the wetting dynamic is driven by surface tension, viscosity, and temperature for a given liquid–solid system (Sect. 3.2). Figure 2.4 shows the effect of temperature on the surface tension and viscosity of water [14]. Both viscosity and surface tension decrease as temperature increases. Therefore, controlling the temperature of the measuring system will go a long way in getting consistent data. Solvents with high viscosity can still be problematic because wetting may become hydrodynamic control (Sect. 3.2). While it may take water less than a second to reach the static state, the time it takes for a

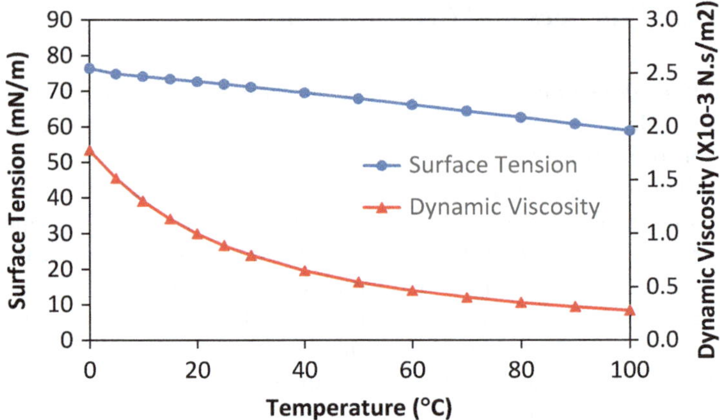

Fig. 2.4 Effect of temperature on surface tension and dynamic viscosity of water, data from [14]

viscous liquid to reach its static state can be minutes to hours (Fig. 3.5). As a recommendation, we would suggest to capture the wetting process with a video from the moment the drop contacts the surface to the point where it reaches the final static state (Fig. 2.3a–c) for viscous liquids or any liquid–solid systems when there is a concern if a static state is attained or not.

According to the fundamental of wetting dynamics, the sessile droplet will reach its static state when all the kinetic energy in the droplet is consumed (Sects. 3.2 and 3.3). In order to obtain a consistent drop profile, one has to have a precise control of the drop size as well as minimizing the variation of the kinetic energy input during the dispensing process. In most commercial goniometers, testing liquid is dispensing through a motor-controlled dosing unit, and the drop volume can be controlled within a percent. There are many ways to form a sessile droplet on the solid surface. For example, one can generate a liquid droplet and suspend it at the tip of the needle first. The dispensing process can be (a) dropping the liquid droplet onto the solid surface through gravity, or (b) moving the sample table up very slowly and allow the droplet to be "picked up" by the surface from the needle. Each method has its advantage and disadvantage. For example, dispensing the drop through gravity can eliminate any influence of the needle on the drop shape. The downside is that even the drop height is tightly controlled and to a minimum drop height, the drop nevertheless gains a small amount of kinetic energy. Depending on the liquid and the surface, varying the drop height can generate a different wetting state and a different contact angle. Two practical examples illustrating the drop height effect on contact angle are discussed in Sects. 3.6 and 4.3.3.

The potential influence of the drop profile by the needle always exists for the drop pickup method. While this method minimizes the kinetic energy gained by the drop during gravity dispensing, it is usually difficult to pick up a suspended drop onto a solid surface with extremely high repellency due to the strong adhesion between the drop and the dosing needle as compared to the wetting surface. Smaller needles with a Teflon-like coating at the tip can facilitate liquid dispensing in this situation. If the suspended liquid drop cannot be easily transferred onto the solid surface, another option is to keep the sample stage in the measurement position, adjust the distance between the dosing needle and the sample so that at the end of the dosing process, the drop will touch the sample surface and a rapid drop deposition on the sample is resulted. The choice of method may depend on the liquid–surface system one is investigating. Sometimes, careful planning and pretesting may be needed to determine the proper dispensing method for a given study.

2.2.3 Effect of Drop Size

According to the Young's equation, the static contact angle is not a dependence of the drop size. However, it is intuitively expected that gravity will start distorting the shape of the drop when the volume of the liquid become very large, which may affect the contact angle. This issue was recently studied thoroughly by Extrand and Moon [15] who examined the drop shapes and contact angles of water, ethylene glycol, and diiodomethane on a Teflon PFA (perfluoroalkoxy copolymer) surface as a function of the volume of the sessile droplet. Assuming the drop is spherical, the sessile droplet can be described by its base diameter ($2a$), height (h), and contact angle (θ) (Fig. 2.5a). The effect of drop volume as it varies from 1 to 2000 μL on $2a$, h and θ was studied for the three solvents. Results show that both $2a$ and h increase initially as the drop volume increases in the small volume regime and h tends to level off for volume >1000 μL. The results for water are plotted in Fig. 2.5b, c. When $2a$ is plotted against h, one can clearly observe the existence of three volume regimes. In the small volume regime (less than ~10 μL), $2a$ is linearly correlating to h indicating that the drop expands spherically as the volume is increasing. In the large volume regime (>1000–2000 μL), increasing the volume of the drop does not lead to any increase in drop height. This suggests that for large drop, the shape is distorted and controlled by gravity. In between is the intermediate regime. Extrand and Moon also measured the transition points when gravity starts having an effect. The transition volumes for water, ethylene glycol, and diiodomethane were found to be at 39, 25, and 6.3 μL, respectively. According to theoretical modeling, the decrease of surface tension and increase of density are the deciding factors for the decrease of transition volumes for ethylene glycol and diiodomethane. The sessile drops and contact angle data for water on PFA with drop sizes of 1, 50, and 2000 μL are given in Fig. 2.5d–f, respectively. The angles were curve-fitted with the tangent method and are found to be the same within experimental error. Similarly, Kranias [16] showed the absence of any drop volume effect for drops ranging from 1 to 10 μL on both hydrophilic and hydrophobic surfaces. As a preference, Kranias indicated that smaller drops are preferred for hydrophilic surfaces, while larger drops are more desirable for hydrophobic surfaces.

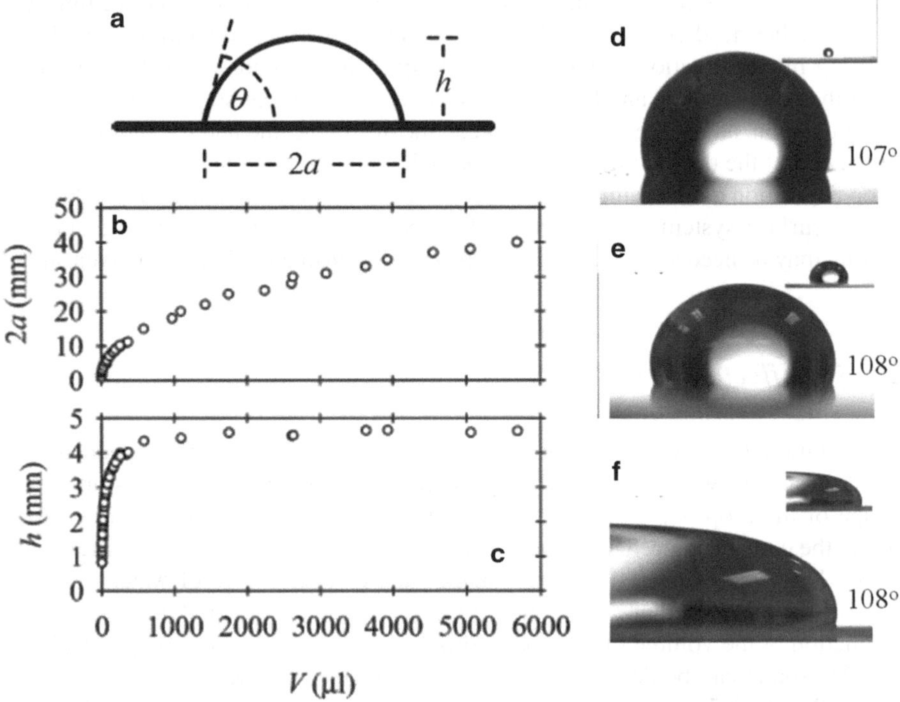

Fig. 2.5 (a) Dimensional parameters for a sessile drop, (b, c) plots of drop diameter and height as a function of drop volume for water on PFA, and (d, e, f) images of a 1, 50, and 2000 μL water sessile drop on PFA [inset: photographs of the same drops at the same magnification] (Reproduced with permission from [15], Copyright 2010 The American Chemical Society)

For contact angle measurement on surfaces with large contact angles, which are usually rough or textured, care should be taken in selecting the drop size for accurate, reproducible measurement. One may have to identify a drop size that will be at least 2–3 orders of magnitude larger than the roughness length scale to avoid significant distortion of the contact line due to roughness [6, 17], at the same time, small enough to make sure that the drop shape will not be overly distorted by gravity. Gravity is known to distort the shape of sessile droplet with large contact angle (>150°) to a larger extent due to the small contact area underneath the drop [18–20].

Extrand and Moon [18] reported the use of the classic numerical solutions from the Bashforth and Adams equation [21] to study the role of flattening of drops by gravity on superhydrophobic surfaces. The calculated theoretical drop profiles agree well with the experimental drop profile on a completely non-wetting TFE (tetrafluoroethylene oligomers)-modified surface. They concluded that the shape and any perceived additional spreading (or wetting) for the larger drops are due to distortion by gravity. Localized distortion near the bottom portion of the water drop was

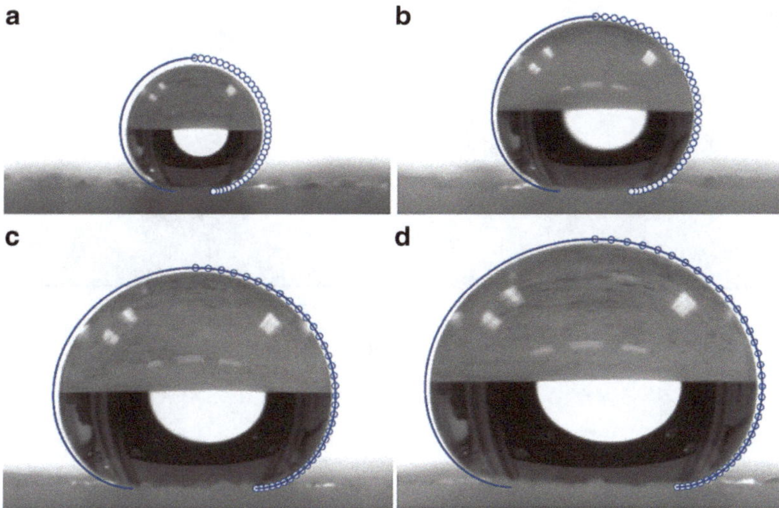

Fig. 2.6 Images of sessile water drops (**a**) 3.4 µL, (**b**) 8.7 µL, (**c**) 20.7 µL, and (**d**) 33.2 µL on a completely nonwetting hydrophobic surface made from TFE powder. The curves are theoretical profiles calculated from the Bashforth and Adams equation (Reproduced with permission from [18], Copyright 2010 The American Chemical Society)

observed for the 3.4 µL droplet, and the distortion is shown to increase as the drop size increases and becomes more global for the 20–33 µL drops (Fig. 2.6).

For low surface tension and high density liquids, this flattening effect due to gravity is more pronounced [18, 19]. Figure 2.7b–d show the sessile drop images of the diiodomethane, ethylene glycol, and hexadecane droplets. The distortion of these droplets, while smaller in drop sizes, is comparable to that of the distorted water droplet at 33.2 µL (Fig. 2.7a).

On the one hand, smaller drops are sensitive to optical errors associated with light scattering, diffraction, evaporation, and correspondingly uncertainty in precisely locating the baseline and digitizing the drop profile. In addition, line tension effect on contact angle measurement may need to be considered when extremely small drops are studied [22, 23]. We would recommend the use of smaller drops when measuring the contact angles of super repellent surfaces (>150°). The choice of the drop size may be a balance between the gravity effect and the roughness length scale effect on the drop distortion.

A different observation was reported by Cansoy [24] who showed that the measured contact angles for water on several microtextured superhydrophobic surfaces are insensitive to the drop size ranging from 0.5–19 µL. There may be multiple causes for the observation, such as dispensing procedure, extraction of baseline from the drop profile or drop shape analysis. The discrepancy between the Cansoy observation and early work may be resolved if details of the dispensing procedure is provided and images of the sessile droplets and procedure for the drop shape analysis are reported.

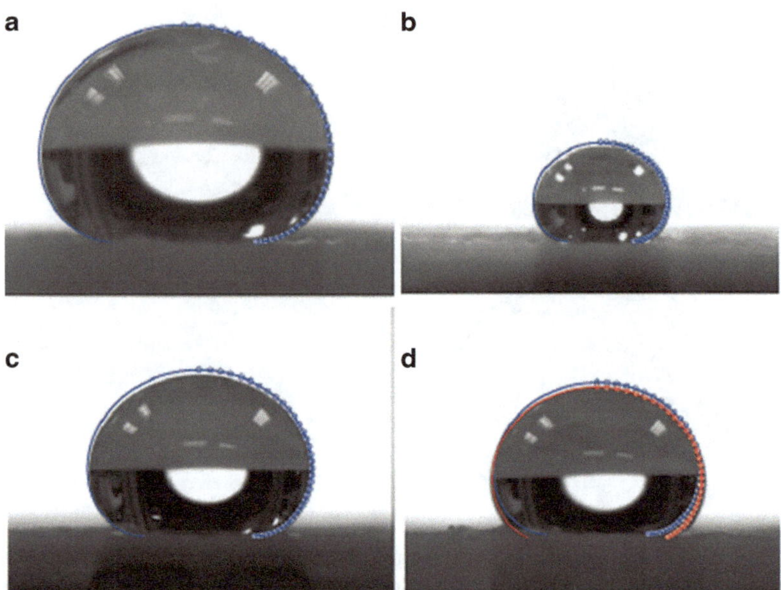

Fig. 2.7 Images of sessile droplets from (**a**) water 33.2 μL, (**b**) diiodomethane 3.2 μL, (**c**) ethylene glycol 15.2 μL, and (**d**) hexadecane 11.6 μL. The curves are theoretical profiles calculated from the Bashforth and Adams equation (Reproduced with permission from [18], Copyright 2010 The American Chemical Society)

2.2.4 *Drop Image Capturing and Drop Shape Analysis*

The repeatability and accuracy of contact angle measurements, to a large extent, depend on the consistency of capturing the drop profile, establishing the baseline on the sessile drop profile, and performing accurate drop shape analysis. In the goniometer, the lens and the CCD camera are preferably tilted 1–3° down towards the horizontal plane to avoid blocking the contact line of the sessile drop [25]. This arrangement also brings portion of the drop profile near the liquid–solid interface into focus and includes a small amount of reflection image below the baseline which greatly facilitates the detection of the baseline. Typical practices include light brightness adjustment, selectively blocking part of the light, and turning off other room lights if possible. A clear and sharp liquid–solid interface, high-quality sessile drop image with sharp and focused boundary, will reduce errors in assigning the baseline and fitting the drop profile.

After the drop shape profile is captured, the contact angle can be calculated using curve-fitting software available in most of the commercial goniometers. Several mathematical methods have been used to calculate the contact angles at the three-phase contact line, including tangential method, $\theta/2$ method and circle method [26–28], ellipse method [29, 30], Young–Laplace method [31–35], polynomial method [36, 37], and B-spline snakes method [38].

Fig. 2.8 Schematic of the $\theta/2$ method of a sessile drop

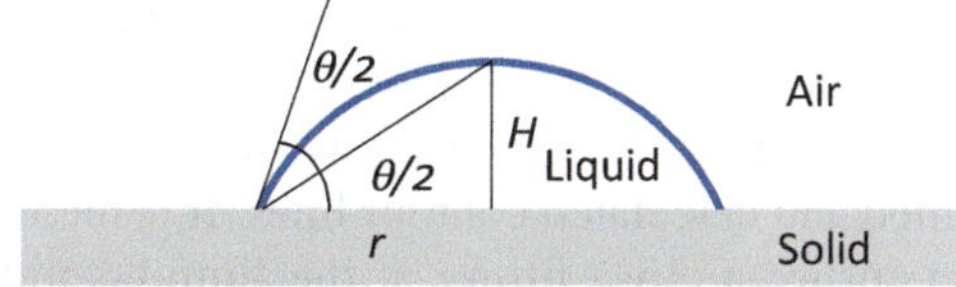

Tangential method is to directly take the tangent at the contact point after digitization of the drop profile. This method is a curve-fitting method and may have large errors due to disturbance in the drop shape caused by dust particles or contaminants or surface irregularities.

$$\theta = \tan^{-1}\left(\frac{dy}{dx}\right)\Big|_{\text{contact point}} \tag{2.2}$$

$\theta/2$ method and circle method [26–28] assume that the liquid drop on the sample surface is part of a sphere under the effect of liquid surface tension only (Fig. 2.8). $\theta/2$ method calculates the contact angle by measuring the drop height H and the contact base radius r:

$$\theta = 2 \cdot \tan^{-1}\left(H/r\right) \tag{2.3}$$

Circle method performs a circle fitting for the detected drop profile. The static contact angle is calculated between the baseline and the tangent of the fitted circle at the contact points, which are the two intersection points of the baseline and the fitted drop profile. $\theta/2$ method and circle fitting method are suitable for surfaces with small contact angles or when very small drop volumes are used because they assume no gravitation effect on the drop shape.

Ellipse method [29, 30] assumes that the liquid drop on the sample surface is part of an elliptical shape. An ellipse equation is fitted to the drop profile to generate least fitting error. The static contact angle is calculated at the contact points based on the fitted ellipse equation and the baseline. However, ellipse fitting method is using mathematic algorithm without physical meanings. This may lead to large deviation between the fitted curve and the captured drop profile, especially for large droplets with large contact angles.

Young–Laplace method provides excellent repeatability if a sessile drop presents a high degree of symmetry in its drop shape. It is also called axisymmetric drop shape analysis (ADSA). It assumes that gravity is the only external force, and the drop shape is axisymmetric under the surface tension and gravity forces. After Rotenberg et al. [31] creates the first generation of ADSA algorithm, significant improvements have been made to enhance the robustness of the method, including ADSA-P (profile) [32], ADSA-D (diameter) [33, 34], and ADSA-NA (non-apex) [35]. Among the ADSA methods, ADSA-P is the most frequently used algorithm in the commercial goniometer software. The ADSA methods adjust the interfacial surface tension and solve the Laplace equation iteratively to produce theoretical drop profiles with least fitting error to the experimental drop profiles. This method has been used extensively in pendant drop analysis. In contact angle measurement, it is also often used if the drops are highly axisymmetric.

The fitting error shows how well the captured drop contour profile agrees to the fitted and calculated contour lines. It is recommended to check the total fitting error to ensure a close fitting at the liquid–solid–air three-phase interface. Recently, Xu [39] reported an error analysis of the three fitting methods (circle fitting, ellipse fitting, and ADSA-P fitting) on water drop profiles with different drop volumes, contact angles, and noise levels. The errors in the contact angles by the circle and ellipse fitting increase as the drop volume increases. Circle fitting has a larger fitting error than that of the ellipse fitting method. The deviation occurs when the water drop profile deviates from the circle or ellipse assumption as drop volume increases. The results also showed that the contact angle error increases as the contact angle becomes larger for the circle and ellipse fitting methods. Based on this analysis, Xu [39] defined a critical water drop volume which corresponds to the water drop volume with a certain contact angle error. Presumably, if the water drop volume is less than the critical water drop volume, the fitting error will be less than the specified contact angle error. The Xu paper is worth reading for anyone who is interested in learning more about drop size and curve-fitting errors.

The ADSA-P method gives the lowest fitting error as both the drop volume and contact angle increases. However, it gives rise to greater errors when there is high level of noise in the drop profile and when the drop shape is deviated from a perfect axisymmetric shape [39]. For very small contact angles (e.g., <15°), circle fitting gives the most robust fitting against noise. Ellipse fitting provides very good robustness against noise except very small and very large contact angles. Therefore, for a typical surface with contact angle ranging between 20° and 120°, ellipse fitting of a 5 µL drop can be safely used to calculate the surface contact angle [30, 39]. Generally speaking, circular fitting will provide minimal fitting error for contact angle smaller than 20°. As for contact angle larger than 150°, a combination of using a small drop size and the ADSA fitting software may produce the most desirable result.

For non-axisymmetric drop shapes, circle fitting and the ADSA fitting cannot be performed without resulting in large fitting errors. Although ellipse fitting was employed to calculate the contact angles of non-axisymmetric drops in some special cases [40], it is not capable of fitting any shape of the drop especially when surface wetting gradient and surface heterogeneity exist. Polynomial fitting and B-spline fitting can be employed to fit the drop profile above the baseline, without any assumption on the drop shape. Therefore, these two methods can be applied in various conditions, large or small drops, symmetric or asymmetric drops, and various contact angles, etc. By optimizing the fitting parameters, this method can provide closer fitting to the drop profile.

Polynomial fitting [36, 37] is different from the ADSA global fitting. It fits the captured drop profile near the local contact point, and the contact angle is calculated using the slope of the polynomial at the contact points to show the local surface wetting property. The order of the polynomial and the number of pixels in the curve-fitting procedure are the two main parameters for polynomial fitting. Bateni et al. [37] recommended a third-order polynomial with 120–140 pixels as an appropriate combination for measuring water contact angle on PMMA surface (72°–74°). However, fitting error analysis should be performed to find out the optimum fitting parameters (e.g., polynomial order and number of pixels) for a specific application. Higher order polynomials, which require more number of pixels for the fitting, are not necessarily better since they are more sensitive to experimental noises.

B-Spline fitting [38] defines the contour of the drop as a versatile B-spline curve without making physical assumptions. Similar to polynomial fitting, B-spline fitting provides contact angles at the position of the contact points; but it also retains the global fitting of the drop shape. This is especially helpful to precisely detect the liquid–solid interface and contact points using the reflection of the drop. This method has been implemented in the free software Image-J as DropSnake fitting method. Inter-knot distance is a critical parameter, which determines how many knots are needed to correctly represent the drop profile. Due to its elasticity of the active contour, it is very flexible for all different kinds of contact angle measurements, static or dynamic contact angles, nonsymmetric or even irregular drop shapes.

It is worth noting that the polynomial and B-spline fitting methods are sensitive to some objective selections of fitting parameters, for example, fitting orders and the number of pixels for fitting, inter-knot distance. In addition, disturbance in the drop shape caused by dust particles or contaminants or surface irregularities may also affect the accuracy of the contact angle measurements.

Selection of curve-fitting method appears to be important in determining the contact angles for super repellent surfaces. With the same 5 μL water droplet, Zhang et al. [19] showed that the calculated contact angles can vary from 152° to 179.8°, depending on the curve-fitting method used (Fig. 2.9). Tangential fitting and circle fitting gave a contact angle of 152–153°, the lowest among the four fitting methods. A contact angle of 156.5° was obtained by ellipse fitting, while a 179.8° contact angle was calculated from the Laplace–Young fitting. By comparing the fitting curves and the drop profile, large deviations can be observed near the contact lines for the circle fitting, tangential fitting as well as the Laplace–Young fitting. The ellipse fitting seems to render the smallest fitting error, presumably producing the most accurate contact angle measurement.

However, the ellipse fitting may have shortfall for large droplets on super repellency surfaces due to flattening of the drop by gravity [18, 19]. Figure 2.10a shows the drop profile of a 5 mg water drop on a superhydrophobic surface with a measured contact angle of 156° using ellipse fitting [41]. After 40 min exposure in air, the drop volume reduces to 0.3 μL and drop distortion (by gravity) is visibly reduced. The contact angle is calculated to be 173° by ellipse fitting. The discrepancy in contact angles between different fitting methods can be minimized when using smaller drops [18].

In 2011, Srinivasan et al. [20] systematically assessed the contact angle measurement of superhydrophobic surfaces using a perturbation solution of the Bashforth-Adams equation. The uncertainty in the calculated contact angle is determined by the uncertainty in the drop height measurement. For example, assuming the resolution limit of the imaging camera is 10 μm, for a 0.7 μL water drop of 153° contact angle calculated from Bashforth-Adams equation, will result in an uncertainty of 3° in the

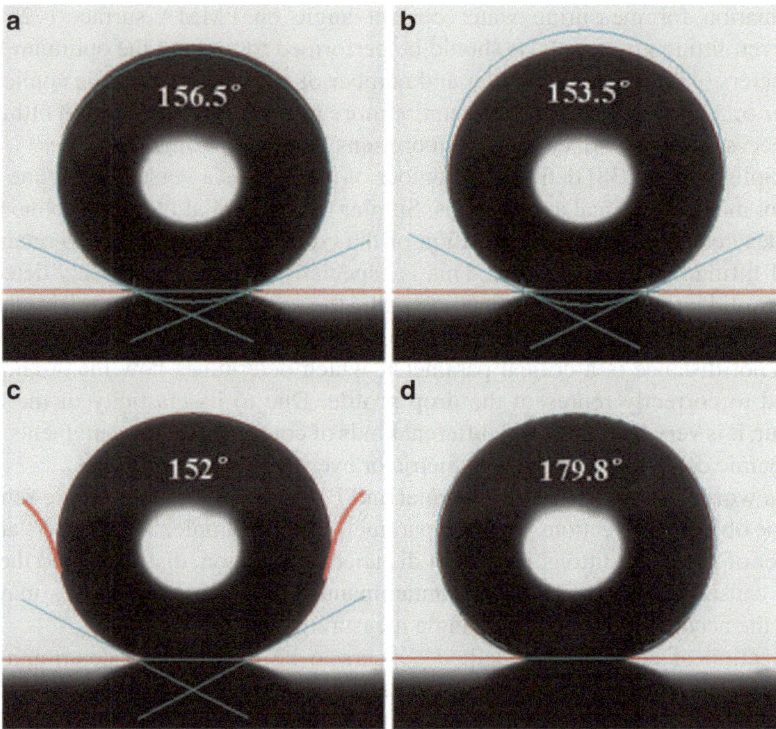

Fig. 2.9 Comparison of curve-fitting methods for a 5 μL water sessile drop on a superhydrophobic surface: (**a**) ellipse fitting, (**b**) circle fitting, (**c**) tangent searching, and (**d**) Laplace–Young fitting. The figures include the simulation lines of the shape of the water droplets and the horizontal baselines (Reproduced with permission from [19], Copyright 2008 The Royal Society of Chemistry)

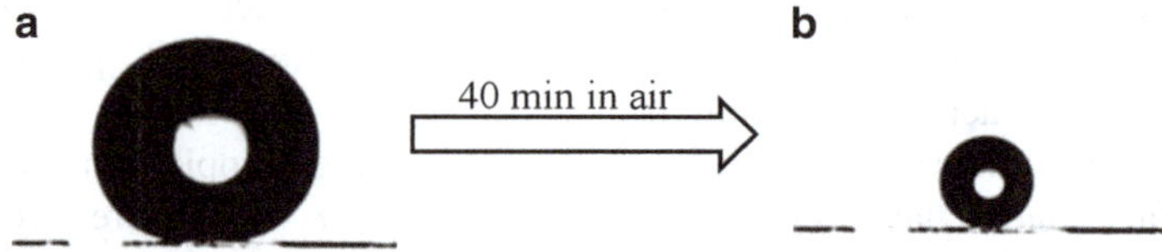

Fig. 2.10 (**a**) Image of a 5 mg water sessile droplet on a superhydrophobic surface and (**b**) drop in (**a**) exposed in air for 40 min (Modified with permission from [41], Copyright 2004 The American Chemical Society)

contact angle. On a hypothetical surface with a true contact angle of 179°, an uncertainty as small as 1 μm in the measurement of the height or the location of the baseline for a 1.4 μL water drop can lead to an uncertainty of 10° in the contact angle. A similar concern exists in contact angle sensitivity for various sessile drop fitting methods, when contact angles approach 180°, the uncertainty in contact angle measurements increases dramatically due to the imaging resolution in height or in defining the location of the baseline. It is desirable to have high quality optics and cameras to enable accurate capture of the drop shape on super repellent surfaces.

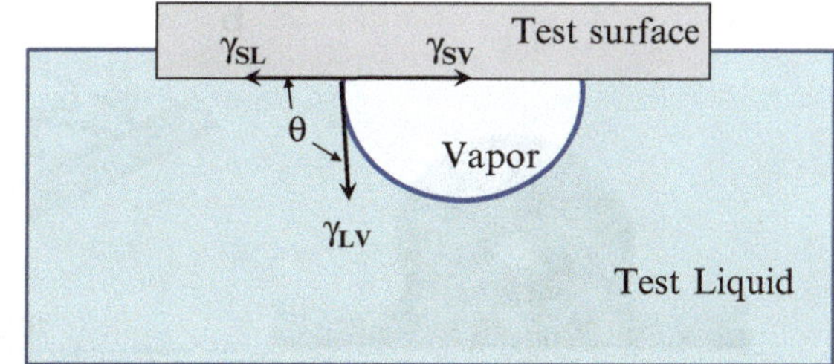

Fig. 2.11 Schematic of contact angle measurement using the captive bubble method

2.2.5 Captive Bubble Method

An alternate method frequently mentioned in the literature is the so-called captive bubble technique [42]. Experimentally, an air bubble of known volume is produced at the tip of a microsyringe. This bubble is then injected into a tank containing the test liquid. The test surface is positioned on the top of the test liquid and 2–3 mm above the injected air bubble. A captive bubble is formed when the air bubble floats upward and be "captured" by test surface. Contact angle θ, formed at the three phase contact line, follows the same Young's equation in Eq. (2.1) and can be calculated from the drop profile of the bubble in a manner similar to the sessile drop method. A schematic of the captive bubble method is shown in Fig. 2.11.

Comparing to the sessile drop method, the captive bubble method has the advantage of measuring contact angles without the influence from the needle and drop height. Temperature can be easily controlled through the liquid, and the vapor phase is always in equilibrium with the liquid phase. Disadvantages are also obvious. A large quantity of liquid has to be used and the solid is in contact with the test liquids over long period of time, which may lead to swelling or other side effect. Although a good agreement was achieved on clean smooth surfaces between the sessile drop and captive bubble methods [43], the latter is still limited to measurement of contact angles on special surfaces. This in part may be due to the simplicity of the sessile-drop method.

2.3 Determination of Sliding Angle

2.3.1 Measurement Procedures and Details

Sliding angle measures surface stickiness or the mobility of the liquid drop on a solid surface. Experimentally, after a sessile drop is formed at the horizontal position, the sample stage together with the solid sample is then tilted at a very low speed. Sliding angle α is defined as the angle when the drop starts to move on the tilted surface. The tilting speed needs to be low enough, e.g., ~1°/s, to ensure that the motion of tilting will not cause any drop movement. Figure 2.12a shows a photograph of a water drop on a titled surface. Clearly, the spherical droplet is distorted

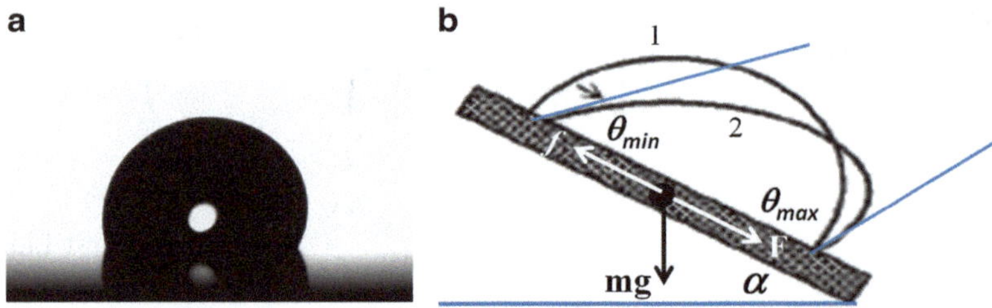

Fig. 2.12 (**a**) A photograph of a water droplet on a tilted surface and (**b**) a 2D schematic showing the effect of tilting on the profile of the sessile droplet

by gravity due to tilting and the 2D view of the sessile drop showing the tilting action is depicted in Fig. 2.12b. When the surface is tilted, a gravitational force is acting on the drop. This force pulls the drop downward, causing distortion of the drop shape, e.g., from a spherical shape to the shape shown as curve 1 schematically illustrated in Fig. 2.12b. The pull force F is given by:

$$F = mg \cdot \sin \alpha \qquad (2.4)$$

where m is the mass of the drop, g is gravitational constant, and α is the tilt angle.

The retention force or friction that keeps the drop from sliding is f and is given by [44]:

$$f = \gamma_{LV} \cdot R \cdot k \cdot \left(\cos \theta_{min} - \cos \theta_{max} \right) \qquad (2.5)$$

where γ_{LV} is the surface tension of the liquid, R is the length scale for the contour of the drop, k is an adjustable parameter based on experimental data, θ_{max} is the contact angle at the lead edge, and θ_{min} is the contact angle at the trail edge.

As the tilt angle continues to increase, the drop shape is distorted further, curve 1 to 2 in Fig. 2.12b. When $F = f$, the drop begins to move and the tilt angle is recorded as α.

The criteria to determine if a drop starts to move is very subjective. Since the drop will only start to slide when the contact line at the trail edge is de-pinned, it would be very helpful to place a tiny droplet of a nonvolatile liquid next to the trial edge as a positional reference. The volume of the reference drop is so small that it will never move during tilting. By carefully examining the distance between the reference drop and the trail edge, one can easily capture the moment when the drop starts to slide.

Since the starting point for the sliding angle measurement is the sessile drop on a horizontal surface, for consistency and reproducible measurement, all cautious details discussed previously for preparing the sessile drop are applied. This will ensure the same metastable wetting state will start the titling experiment. As an antidotal experiment, Pierce, Carmona, and Amirfazli [45] demonstrated that the sliding angles measured for drops prepared on inclined surfaces are consistently smaller than those prepared from leveled surfaces.

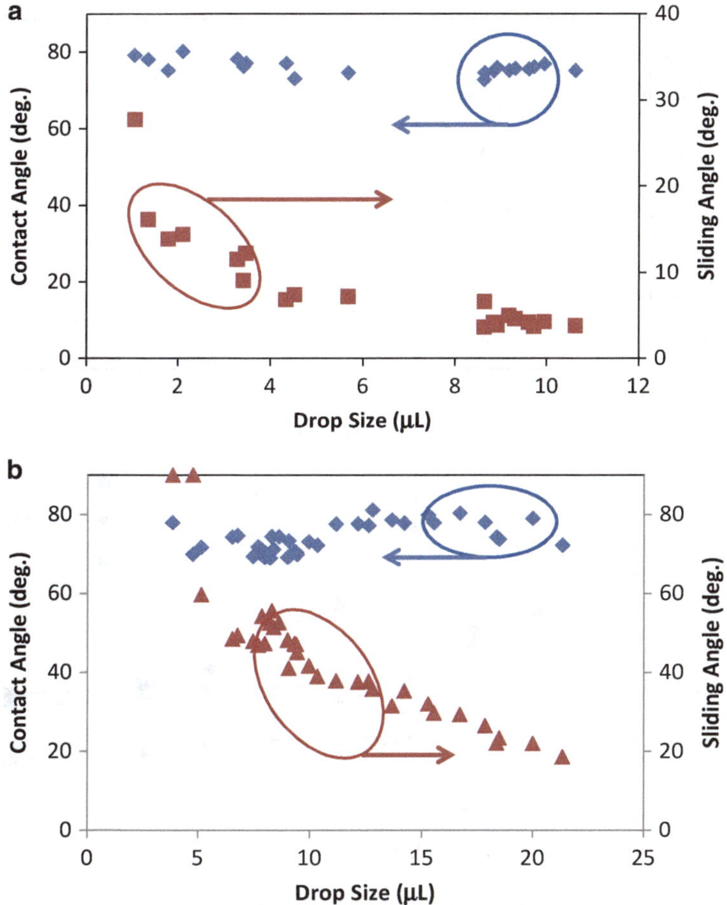

Fig. 2.13 Effect of drop size on the contact angle and sliding angle of hot polyethylene droplets on (**a**) a hydrophobic sol gel coating and (**b**) a fluorinated self-assembled-monolayer on silicon wafer

2.3.2   *Effect of Drop Size*

According to Eq. (2.4), the measured sliding angle is a strong function of the drop mass/volume, the heavier/larger the drop, the smaller the sliding angle. Figure 2.13a, b show plots of the contact angles and sliding angles as a function of the size of hot polyethylene wax droplets on two different surfaces. Consistent with the discussion in Sect. 2.2.3, variation of the drop size has little effect on the contact angle. In contrast, there is a strong drop size effect on the sliding angle, the larger the drop the smaller the sliding angle.

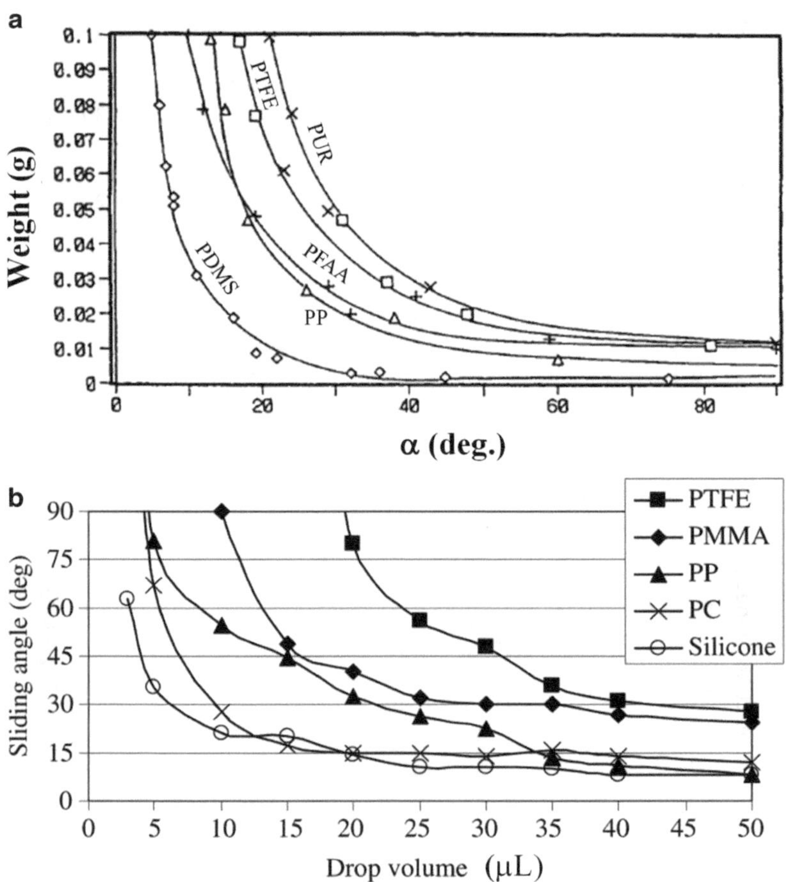

Fig. 2.14 (**a**) Plot of water drop mass versus sliding angle on five different polymer surfaces (Reproduced with permission from [46], Copyright 1994 John Wiley & Sons), and (**b**) plot of water sliding angle versus drop volume on polymer surfaces (Reproduced with permission from [47], Copyright 2007 Taylor & Francis Group)

The drop size dependent on sliding angles is very general and has also been observed for water and hexadecane on other surfaces. In 1994, Murase et al. [46] reported a study of the interaction between water and 20 different polymer surfaces. They also observed a strong drop mass effect on the sliding angle, and a typical plot is highlighted in Fig. 2.14a. Later Rios and co-workers [47] reported the use of the sliding angle to probe water/ice adhesion on five different polymers and also observed a strong drop size effect. In view of the strong mass/volume dependence, we recommend that drop mass/volume has to be reported along with the sliding angle data. It would be misleading if one claims a surface sticky for a 1 μL drop that is not sliding at 90° tilt angle or a fast sliding drop with drop volume of > 20 μL for example. The data in Figs. 2.13 and 2.14 suggest that the appropriate range to study sliding angle would be ~5–10 µL depending on the liquid–surface system being investigated. This is the range where the sliding angle tends to be most sensitive when the drop mass/volume is varying.

2.4   Determination of Advancing and Receding Contact Angle

2.4.1   Needle-embedded Sessile Drop Method

Advancing and receding contact angle have been known for more than a century to be the maximum and minimum angle that can be measured reliably and reproducibly when liquid is added to or withdrawn from a liquid droplet at a very slow rate [4, 5]. Most of the current goniometer is equipped with an automated dosing system, where drop volume and rate of addition/withdraw can be controlled very precisely. The needle-embedded sessile drop method is the most frequently used method and is schematically shown in Fig. 2.15.

Experimentally, a small sessile drop (1–2 µL) is first dispensed on the test surface, small amount of liquid is then precisely pumped through the microsyringe to the sessile drop at a rate of <0.2 µL/s. As the volume of the sessile drop is increased, the contact line advances outward. The advancing experiment is usually considered as complete when the drop size increases to ~20 µL. It is a good practice to let the system stabilized for a few seconds prior to the receding angle measurement, which is done by reversing the springe pump and withdrawing the liquid from the expanded sessile drop at the same slow rate. Images of the drop profiles during advancing and receding are captured and an example of the output is given in Fig. 2.16. From the drop profiles, the advancing and receding contact angles at different drop volumes can be calculated used curve-fitting software as described in Sect. 2.2.4. Usable fitting methods include: tangential method, ellipse fitting, polynomial fitting, and B-spline fitting. Tangential method is usually preferred for receding contact angle calculation especially if the drop changes into a triangle shape during the receding

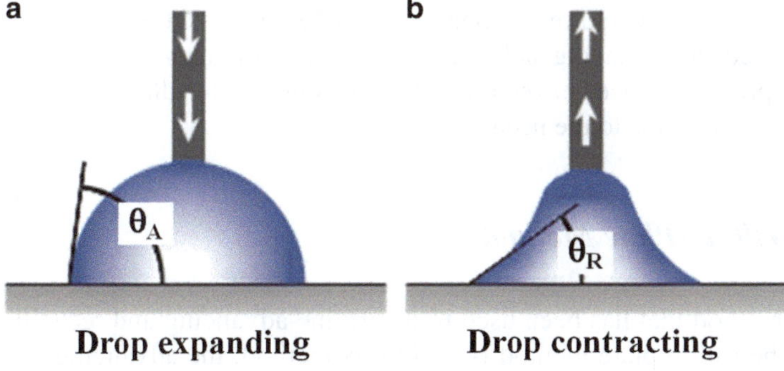

Fig. 2.15 Schematic showing measurement of θ_A and θ_R using the drop expansion/contraction technique

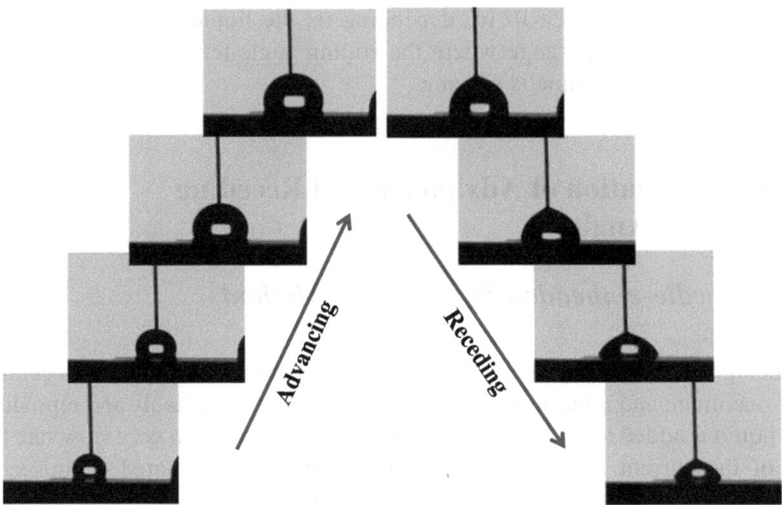

Fig. 2.16 Sessile drop images of water droplets on PTFE during advancing and receding contact angle measurement

process. Similar to static contact angle measurement, one should check the fitting error to ensure a close fitting especially near the contact points region.

Since the needle is embedded in the sessile drop, the ADSA-P method will not be applicable due to the missing apex on the drop profile and the embedded needle. However, there are two approaches to apply the Young–Laplace method to calculate the advancing and receding angles. One approach is to dispense the liquid through a hole in the bottom of the surface [45]. Carefully prepared solid surface will ensure axisymmetric drop shape during drop expansion and contraction. Another approach is to use the recently developed ADSA-NA (no-apex) drop shape analysis algorithm [35], which does not require information about the apex region.

Dosing of low surface tension liquids during the advancing contact angle experiment may be problematic due to the strong adhesion between the small diameter needle and the liquid. While switching to a larger diameter needle may circumvent the problem, larger diameter needles however tend to produce more disturbance. If that is the case, we would recommend using a Teflon coated dosing needle for the needle-embedded advancing and receding angle measurements. This approach may provide a proper balance between liquid dispensing due to adhesion issue and drop shape disturbance due to the needle.

2.4.2 Tilting Plate Method

Another method that has been used to determine advancing and receding contact angle is the tilting plate method [48]. In this method, the advancing angle is the contact angle at the lead edge, and the receding angle is the contact angle at the trail edge of the distorted droplet on the inclined surface when the drop starts sliding (Fig. 2.17).

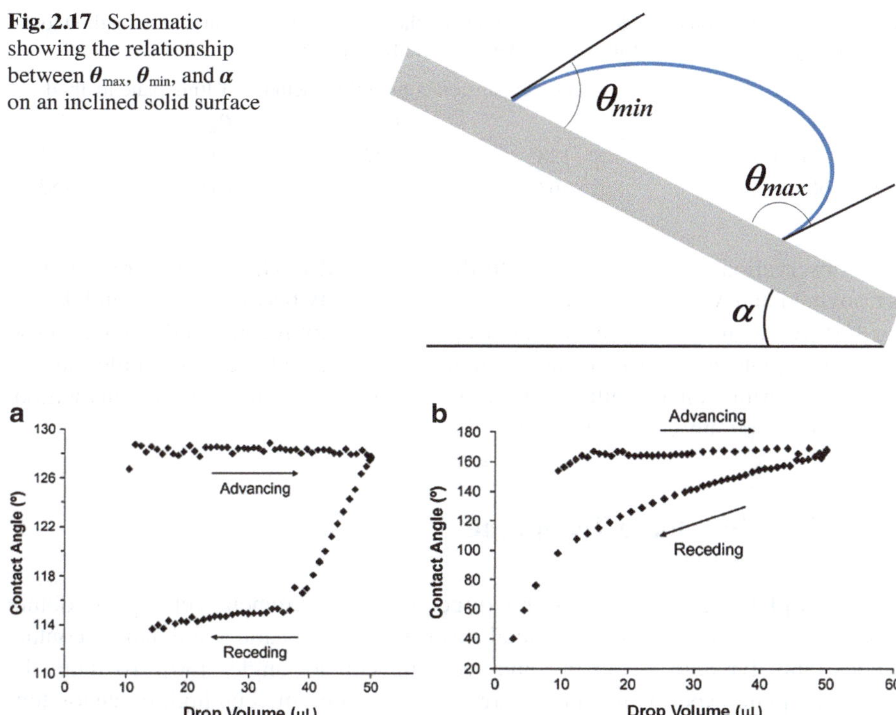

Fig. 2.17 Schematic showing the relationship between θ_{max}, θ_{min}, and α on an inclined solid surface

Fig. 2.18 Advancing and receding contact angle measurement for water droplets on (**a**) Teflon coated silicon and (**b**) AKD polymer surfaces with the drop expansion/contraction technique (Reproduced with permission from [45], Copyright 2008 Elsevier)

When the drop just about to slide at α, $F=f$. From Eqs. (2.4) and (2.5), plot of $\sin\alpha$ versus ($\cos\theta_R-\cos\theta_A$) should yield a linear relationship if $\theta_A=\theta_{max}$ and $\theta_R=\theta_{min}$. Experimentally, a very scattered plot was obtained (Fig. 5.13), indicating that $\theta_A\neq\theta_{max}$ and $\theta_R\neq\theta_{min}$. The fundamental related to the inequality between θ_A/θ_{max} and θ_R/θ_{min} will be addressed in Sect. 5.3. In any event, the inequality suggests that the use of the titling plate method to determine sliding angle is technically incorrect.

More recently, Pierce, Carmona, and Amirfazli [45] compared the θ_A and θ_R values of water droplets obtained from the drop expansion/contraction method to the θ_{max} and θ_{min} values obtained from the tilting plate method on a Teflon coated silicon surface and an alkyl ketene dimer (AKD) polymer surface. The advancing and receding contact angle data for the two surfaces are summarized in Fig. 2.18, and comparisons of the angles obtained from these two methods are tabulated in Table 2.1. On the Teflon coated silicon surface, both methods produce similar results. This cannot be said on the AKD surface. Although the θ_A and θ_{max} values are same, the θ_R and θ_{min} values are significantly different. This can be attributed to the low receding angle on the AKD surface. As seen in the data in Fig. 2.18b, due to the high adhesion between water and the AKD surface, the receding contact line was never in mechanical equilibrium during the receding angle measurement.

Table 2.1 Comparison of the θ_A and θ_R values from the drop expansion/contraction method to the θ_{max} and θ_{min} values from the tilting plate method (data from [45])

	Drop expansion/contraction method		Tilting plate method	
	θ_A	θ_R	θ_{max}	θ_{min}
Teflon coated silicon	~128°	~115°	~124°	~112°
AKD polymer surface	~163°	<10°	~163°	~83°

This observation is in agreement with the theoretical modeling study reported by Krasovitski and Marmur [49], who predicted inequality between θ_A/θ_{max} and θ_R/θ_{min} for surfaces with large hysteresis. Hence, we would not recommend anyone to use the tilting plate method to measure their advancing and receding angles unless there is experimental difficulty in executing the drop expansion/contraction method, e.g., high temperature measurement.

2.5 Wilhelmy Plate Technique

Wilhelmy plate method provides an indirect force measurement technique to obtain liquid–air and liquid–liquid interfacial surface tensions, static and dynamic contact angles, and dynamic advancing and receding contact angles [50]. To date, the Wilhelmy plate experiment can be carried out on a commercial high-precision tensiometer, such as model DCAT 21 from DataPhysics. The apparatus consists of a high-precision electrodynamically compensated weighing system with automatic calibrating function with a weighing resolution of 10 μg, software-controlled motorized height positioning of the sample vessel with variable speed ranging from ~0.7 μm/s to ~500 mm/min, and a heating/cooling mantle to control the liquid temperature (−10–130 °C).

In the Wilhelmy plate experiment, the solid surface has to be fabricated in plate form with well-defined dimension both sides. Experimentally, it is immersed into the testing liquid to form a liquid–solid–air interface. The forces exerted on the sample are gravity (g), surface tension ($F_{Wetting}$) and buoyance ($F_{Buoyance}$) forces, and the force by the tensiometer (i.e., action-reaction forces, it is also the force measured by the tensiometer). The free body diagram of the plate is shown in Fig. 2.19 [51]. The equation for the force balance is described as [52]:

$$F_{Tensiometer} = F_{Wetting} - F_{Buoyance} + g \tag{2.6}$$

where the wetting force $F_{Wetting} = \gamma_{LV} \cdot L \cdot \cos \theta$, and L is the wet-length for the three-phase contact line ($2d + 2W$). Usually, the weight of the plate and harness is set to zero on the tensiometer and Eq. (2.6) can be rewritten to:

$$F_{Measured} = F_{Tensiometer} - g$$

$$= F_{Wetting} - F_{Buoyance} = \gamma_{LV} \cdot L \cdot \cos \theta - L \cdot h \cdot \Delta\rho \cdot g \tag{2.7}$$

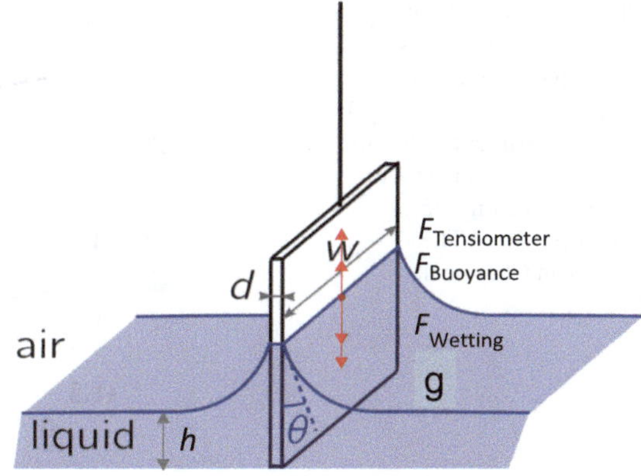

Fig. 2.19 Schematic showing the dimension of the wet-length L and the forces that act on the solid surface in a Wilhelmy plate experiment

where h is the immersion depth, and $\Delta\rho$ is the density difference between the liquid and air.

Static contact angle θ can be calculated from Eq. (2.7) once the force on the tensiometer is recorded, and the geometrical dimensions and immersion depth are measured. Contact angle measured by the Wilhelmy plate method is believed to be more accurate [53, 54]. The method is based on high-precision force measurement and does not have many subjective errors or noise factors, such as needle effect on the drop shape, identification of baseline, and drop shape analysis, associated with the optical sessile drop method. The main disadvantage of the Wilhelmy plate method is its stringent requirements for the test surface. The dimension of the test sample has to be well defined as the wet-length L is a key parameter in calculating the contact angle. In addition, the test sample has to be flat and rigid with homogeneous surface property on both sides.

A significant advantage of the Wilhelmy plate method is its ability to study the kinetics of the moving contact lines in both advancing and receding modes. For instance, instead of measuring the force F_{measured} in the static mode, one can measure it during advancing and receding at a given velocity (V). The wetting force (F_A) during immersion (advancing) as a function of V is given in Eq. (2.8) [52].

$$F_{\text{A}} = L \cdot \gamma_{\text{LV}} \cdot \cos \theta_{\text{A}} \tag{2.8}$$

where γ_{LV} is the surface tension of the liquid, and θ_A is the dynamic advancing contact angle at V.

Similarly the receding force (F_R) withdrawing at a speed of V is given by:

$$F_{\text{R}} = L \cdot \gamma_{\text{LV}} \cdot \cos \theta_{\text{R}} \tag{2.9}$$

where θ_R is the receding contact angle at V.

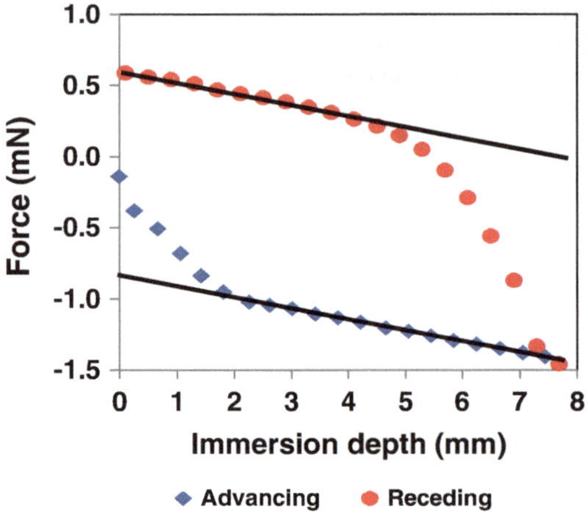

Fig. 2.20 Plot of wetting and receding forces as a function of immersion depth in water at velocity V on an experimental PDMS surface (Reproduced with permission from [55], Copyright 2014 The American Chemical Society)

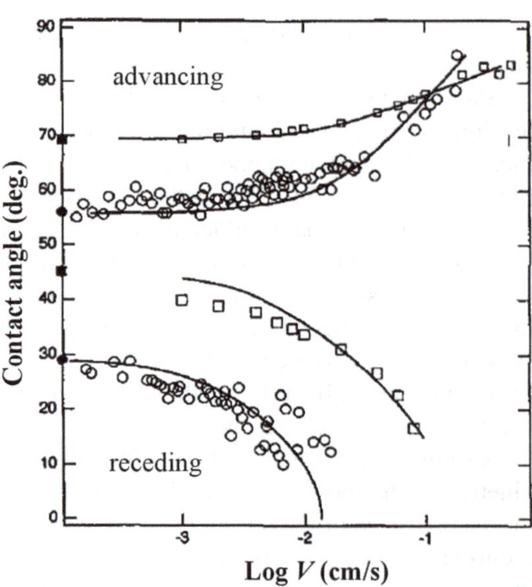

Fig. 2.21 Plot of dynamic contact angles for water (*open square*) and glycerol (*open circle*) on PET as a function of immersion velocity V (Reproduced with permission from [56], Copyright 1993 Elsevier)

Figure 2.20 shows an illustrative example of an experimental plot of F_A and F_R as a function of the immersion depth at V [55]. From the intercepts, the dynamic advancing and receding contact angles (θ_A and θ_R) at V are calculated.

On smooth flat surfaces, the dynamic θ_A/θ_R increases/decreases very slowly in the low-speed regime, and the rate of change increases at higher speed. The θ_A/θ_R values extrapolate to $V=0$ would be comparable to those measured from the drop expansion/contraction method. Typical results can be found in the work of Hayes and Ralston [56], and an illustrative plot is reproduced in Fig. 2.21. In most practical situations, the surface or sometimes the liquid are not static, e.g., liquid movement in pipeline or microfluidic device and printing. The dynamic measurement capability of the Wilhelmy Plate technique would provide a useful tool to study the dynamics of the liquid–solid interactions realistically. In addition, the technique has been shown to be able to quantify contact angle hysteresis, detect roughness length scale as well as wetting transition [55, 57].

Finally, Wilhelmy plate has also been used extensively to measure liquid surface tensions and liquid–liquid interfacial surface tensions by using a roughened platinum/iridium plate rendering excellent wetting by test liquids with a contact angle of ~0°.

2.6 Summary

This chapter summarizes most of the major measurement techniques that are used for surface characterization and wetting studies. In addition to the usual best lab practices, such as clean surface, purified liquid, and ambient condition control, certain specific critical details that may facilitate consistency and enhance reproducibility are also provided. The contact angle measurements include in this chapter are: static contact angle, sliding angle, advancing and receding contact angle, and dynamic advancing and receding contact angle.

Static contact angle has been widely used for surface characterization and is probably the most measured contact angle in the literature. We recommend it to be used in a temperature/humidity-controlled laboratory on a vibration-free table. For data consistency, drop dispense should be gentle with minimal disturbance of the sessile drop by the needle in the microsyringe and drop impact. Close chamber should be employed if volatile liquid is used. Although the Young's angle is not supposed to be sensitive to drop size, we would recommend a 5 µL drop size for general characterization. One may consider using a smaller drop when studying super repellent surfaces, especially with low surface tension and high density liquids. The will eliminate any distortion of the drop profile due to gravity.

Sliding angle is determined by the tilting plate method. The sessile drop should be dispensed horizontally before tilting. Since sliding angle is dependent of drop mass, α must be reported with a drop mass or volume. Large drops, e.g., >20 µL, should be avoided as α becomes less sensitive to mass changes. The proper drop size likely depends on the system studied. Available data suggest that 5–10 µL drops would be appropriate for most systems.

Advancing and receding contact angle should be determined by the drop expansion/contraction technique. The tilting plate method has significant shortfall particularly with high hysteresis samples and should be avoided if possible.

The Wilhelmy Plate technique is the most versatile tool for surface characterization and wetting study. The more elaborate measurement procedure as well as sample preparation have hindered its popularity. However, it is still the best tool when it comes to studying the dynamics of the liquid–solid interactions.

References

1. Young T (1805) An essay on the cohesion of fluids. Phil Trans R Soc London 95:65–87
2. Pease DC (1945) The significance of the contact angle in relation to the solid surface. J Phys Chem 49:107–110
3. Morra M, Occhiello E, Garbossi F (1990) Knowledge about polymer surfaces from contact angle measurements. Adv Colloid Interface Sci 32:79–116
4. Rayleigh L (1890) On the tension of water surfaces, clean and contaminated, investigated by the method of ripples. Phil Mag 30:386–400
5. Bartell FE, Wooley AD (1933) Solid–liquid–Air contact angles and their dependence upon the surface condition of the solid. J Am Chem Soc 55:3518–3527
6. Marmur A, Bittoun E (2009) When Wenzel and Cassie are right: reconciling local and global considerations. Langmuir 25:1277–1281
7. http://www.ramehart.com/newsletters/2007-12_news.htm
8. Bartell FE, Hatch GB (1934) Wetting characteristics of galena. J Phys Chem 39:11–24
9. Zisman WA (1964) Relation of the equilibrium contact angle to liquid and solid constitution. In: Fowkes F (ed) Contact angle, wettability, and adhesion, advances in chemistry. American Chemical Society, Washington, DC, pp 1–51
10. Good RJ (1977) Surface free energy of solids and liquids. Thermodynamics, molecular forces, and structures. J. Colloid Interface Sci 59:398–419
11. Drelich JW (2013) Guidelines to measurements of reproducible contact angles using a sessile-drop technique. Surf Innov 1:248–254
12. Shuttleworth R, Bailey GJ (1948) The spreading of a liquid over a rough surface. Discuss Faraday Soc 3:16–22
13. Zhao H, Law KY, Sambhy V (2011) Fabrication, surface properties and origin of superoleophobicity for a model textured surface. Langmuir 27:5927–5935
14. Crittenden JC, Trussell RR, Hand DW, Howe KJ, Tchobanoglous G (2012) MWH's water treatment: principles and design, 3rd edn. John Wiley & Sons, Hoboken
15. Extrand CW, Moon SI (2010) When sessile drop Are No longer small: transitions from spherical to fully flattened. Langmuir 25:11815–11822
16. Kranias S Effect of drop volume on static contact angles. Technical note 310e. Kruss GmbH, France
17. Marmur A (2006) Soft contact: measurement and interpretation of contact angles. Soft Matter 2:12–17
18. Extrand CW, Moon SI (2010) Contact angles of liquid drops on super hydrophobic surfaces: understanding the role of flattening of drops. Langmuir 26:17090–17099
19. Zhang X, Shi F, Niu J, Jiang Y, Wang Z (2008) Superhydrophobic surfaces: from structural control to functional application. J Mater Chem 18:621–633
20. Srinivasan S, McKinley GH, Cohen RE (2011) Assessing the accuracy of contact angle measurements for sessile drops on liquid-repellent surfaces. Langmuir 27:13582–13589
21. Bashforth F, Adam JC (1883) An attempt to test the theories of capillary action by comparing the theoretical and measured forces of drops of fluid. University Press, Cambridge, UK
22. Amirfazli A, Kwok DY, Gaydos J, Neumann AW (1998) Line tension measurements through drop size dependence of contact angle. J Colloid Interface Sci 205:1–11
23. Marmur A (1997) Line tension and the intrinsic contact angle in solid–liquid–fluid systems. J Colloid Interface Sci 186:462–466
24. Cansoy CE (2014) The effect of drop size on contract angle measurements of superhydrophobic surfaces. RSC Adv 4:1197–1203
25. Woodward RP contact angle measurements using the drop shape method. http://firsttenangstroms.com/pdfdocs/CAPaper.pdf
26. Mack GL (1935) The determination of contact angles from measurements of the dimensions of small bubbles and drops. I. The spheroidal segment method for acute angles. J Phys Chem 40:159–167

27. Mark GL, Lee DA (1935) The determination of contact angles from measurements of the dimensions of small bubbles and drops II. The sessile drop method for obtuse angles. J Phys Chem 40:169–176

28. Neumann AW, Good RJ (1979) Techniques of measuring contact angles. Springer, Surface and Colloid Science, pp 31–91

29. Sklodowska A, Wozniak M, Matlakowska R (1999) The method of contact angle measurements and estimation of work of adhesion in bioleaching of metals. Biol Proc Online 1:114–121

30. Lubarda VA, Talke KA (2011) Analysis of the equilibrium droplet shape based on an ellipsoidal droplet model. Langmuir 27:10705–10713

31. Rotenberg Y, Boruvka L, Neumann AW (1983) Determination of surface tension and contact angle from the shapes of axisymmetric fluid interfaces. J Colloid Interface Sci 93:169–183

32. del Rio OI, Neumann AW (1997) Axisymmetric drop shape analysis: computational methods for the measurement of interfacial properties from the shape and dimensions of pendant and sessile drops. J Colloid Interface Sci 196:136–147

33. Skinner FK, Rotenberg Y, Neumann AW (1989) Contact angle measurements from the contact diameter of sessile drops by means of a modified axisymmetric drop shape analysis. J Colloid Interface Sci 130:25–34

34. Moy E, Cheng P, Policova Z, Treppo S, Kwok D, Mack DR, Sherman PM, Neumann AW (1991) Measurement of contact angles from the maximum diameter of Non-wetting drops by means of a modified axisymmetric drop shape analysis. Colloids Surf 58:215–227

35. Kalantarian A, David R, Neumann AW (2009) Methodology for high accuracy contact angle measurement. Langmuir 25:14146–14154

36. del Rio OI, Kwok DY, Wu R, Alvarez JM, Neumann AW (1998) Contact angle measurements by axisymmetric drop shape analysis and an automated polynomial fit program. Colloids and Surfaces A: Physicochem Eng Aspects 143:197–210

37. Bateni A, Susnar SS, Amirfazli A, Neumann AW (2003) A high-accuracy polynomial fitting approach to determine contact angles. Colloids and Surfaces A: Physicochem Eng Aspects 219:215–231

38. Stalder AF, Kulik G, Sage D, Barbieri L, Hoffmann P (2006) A snake-based approach to accurate determination of both contact points and contact angles. Colloids and Surfaces A: Physicochem Eng Aspects 286:92–103

39. Xu ZN (2014) An algorithm for selecting the most accurate protocol for contact angle measurement by drop shape analysis. Rev Sci Instrum 85:125107

40. Xu ZN, Wang SY (2015) A highly accurate dynamic contact angle algorithm for drops on inclined surface based on ellipse-fitting. Rev Sci Instrum 86:025104

41. Zhang X, Shi F, Yu X, Liu H, Fu Y, Wang Z, Jiang J, Li X (2004) Polyelectrolyte multilayer as matrix for electrochemical deposition of gold clusters: toward super-hydrophobic surface. J Am Chem Soc 126:3064–3065

42. Zhang W, Hallstrom B (1990) Membrane characterization using contact angle technique 1. Methodology of the captive bubble technique. Desalination 70:1–12

43. Drelich J, Miller JD, Good RJ (1996) The effect of drop (bubble) size on advancing and receding contact angles for heterogeneous and rough solid surfaces as observed with sessile-drop and captive-bubble techniques. J Colloid Interface Sci 179:37–50

44. Furmidge CGL (1962) Studies at phase interfaces 1. The sliding of liquid drops on solid surfaces and a theory for spray retention. J Colloid Interface Sci 17:309–324

45. Pierce E, Carmona FJ, Amirfazli A (2008) Understanding of sliding and contact angle results in tilted plate experiments. Colloids and Surfaces A: Physicochem Eng Aspects 323:73–82

46. Murase H, Nanishi K, Kogure H, Fujibayashi T, Tamura K, Haruta N (1994) Interaction between heterogeneous surfaces of polymers and water. J Appl Poly Sci 54:2051–2062

47. Rios PF, Dodiuk H, Kenig S, McCarthy S, Dotan A (2007) The effect of polymer surface on the wetting and adhesion of liquid systems. J Adhesion Sci Technol 3–4:227–242

48. Macdougall G, Ockrent C (1942) Surface energy relations in liquid/solid systems. 1. The adhesion of liquids to solids and a new method of determining the surface tension of liquids. Proc R Soc Lond A 180:151–173

49. Krasovitski B, Marmur A (2005) Drops down the hill. Theoretical study of limiting contact angles and the hysteresis range on a tilted plate. Langmuir 21:3881–3885
50. Wilhelmy L (1863) Ueber die Abhängigkeit der Capillaritäts-Constanten des Alkohols von Substanz und Gestalt des benetzten festen Körpers. Ann Phys 119:117–217
51. Wilhelmy Plate, Wikipedia. http://en.wikipedia.org/wiki/Wilhelmy_plate
52. Liakos IL, Newman RC, McAlpine E, Alexander MR (2007) A study of the resistance of SAMs on aluminum to acidic and basic solutions using dynamic contact angle measurements. Langmuir 23:995–999
53. Martin DA, Vogler EA (1991) Immersion depth independent computer analysis of Wilhelmy balance hysteresis curves. Langmuir 7:422–429
54. Lander LM, Siewierski LM, Brittain WJ, Vogler EA (1993) A systematic comparison of contact angle methods. Langmuir 9:2237–2239
55. Kanungo M, Mettu S, Law KY, Daniel S (2014) Effect of roughness geometry on wetting and de-wetting of PDMS surfaces. Langmuir 30:7358–7368
56. Hayes RA, Ralston J (1993) Force liquid movement on low energy surfaces. J Colloid Interface Sci 159:429–438
57. Kleingartner JA, Srinivasan S, Mabry JM, Cohen RE, McKinley GH (2013) Utilizing dynamic tensiometry to quantify contact angle hysteresis and wetting transitions on nonwetting surfaces. Langmuir 29:13396–13406

Chapter 3
Wetting on Flat and Smooth Surfaces

Abstract The Young's contact angle gives a notion of wettability when a liquid wets and spreads on a solid surface. However, it is also a source of arguments and controversies in the literature. The objective of this chapter is to clarify a couple of fundamental issues that have been debated for more than a century. These issues are: (1) is the liquid droplet in thermodynamic or mechanical equilibrium in the static position as described by Young? And (2) what determines the contact angle, the contact line or the contact area? Studies of the dynamics of the wetting process, by molecular kinetic theory, hydrodynamic theory and experiments, reveal that liquid usually spreads after it wets the surface. The action of spreading is retarded by friction due to molecular adhesion and/or roughness at the liquid–solid interface. The liquid droplet will cease to advance and end up in the final static position with a static contact angle θ when all of its kinetic energy is dissipated. Evidence is provided that the liquid droplet is metastable in this static position. When sufficient vibration energy is provided to the liquid droplet, the contact line is shown to de-pin and the droplet is driven to its thermodynamically stable state, resulting in the most stable droplet with an equilibrium contact angle θ_{eq}. Two simple and clever experimental designs from the literature are reviewed. Both experiments involve the fabrication of chemically heterogeneous surfaces comprising millimeter-size spot of material-1 on a surface of material-2. The contact angles for the spot and the bulk surface are very different. Advancing contact angle measurement was performed from the center of the spot. As the contact line is advancing, the contact angle switches from material-1 to material-2 as soon as the advancing contact line reaches the bulk surface. This is despite the fact that the change in surface energy for the contact area is very small. The results clearly demonstrate that contact angle is determined by the energetics around the contact line, not the contact area underneath the liquid droplet. Recent thermodynamic modelling also confirms that contact angle is a one-dimensional phenomenon.

Keywords Wetting • Young's equation • Smooth surfaces • Wetting dynamics • Dynamic contact angle • Metastable wetting state • Equilibrium contact angle • Drop vibration • Static contact angle • Advancing contact angle • Receding contact angle • Most stable contact angle • Contact line • Contact area

© Springer International Publishing Switzerland 2016

K.-Y. Law, H. Zhao, *Surface Wetting*, DOI 10.1007/978-3-319-25214-8_3

3.1   The Young's Equation

In 1805, Thomas Young [1] stated in his legendary paper that an angle of contact is formed when liquid wets a solid surface. He descriptively stated that this angle, which is commonly known as static contact angle (θ) in contemporary literature, is a result of the balance of three forces acting on the liquid droplet. These three forces are: surface tension of the wetting liquid (γ_{LV}), surface tension of the solid surface (γ_{SV}), and the liquid–solid interfacial tension (γ_{SL}). A vector representation showing the three surface tensions acting at the three phase contact line is given in Fig. 3.1.

This has become the famous Young's equation:

$$\gamma_{SV} = \gamma_{LV} \cdot \cos\theta + \gamma_{SL} \tag{3.1}$$

The concept of contact angle gives a notion of wettability. Zisman [2] defined spreading when $\theta = 0°$ and wetting when $\theta \neq 180°$. In other words, partial wetting occurs in most of the cases we encounter.

3.2   Wetting Dynamics on Smooth Surfaces

Shuttleworth and Bailey [3] noted in 1948 in their study of liquid wetting on rough surfaces that the spreading of a liquid droplet on a solid is a complex phenomenon. The final static position of the advancing liquid not only depends on the surface energy of the liquid, the solid, and the liquid–solid interface but also on the roughness of the surface and the manner in which the liquid is placed on the solid. This statement is not only insightful, but also accurate even by today's standard. In this section, we will focus our attention on wetting of smooth surfaces.

The dynamics of wetting and spreading of liquid droplets on solid surfaces has been studied by many research groups both theoretically and experimentally. Large-scale molecular dynamic simulations of liquid drops spreading on solid surfaces with varying liquid–solid interaction have been studied by Bertrand, Blake, and De Coninck [4]. Figure 3.2 shows profiles of the molecular dynamic simulations for four droplets with increasing liquid–solid interaction. At equilibrium, $C_{SL} = D_{SL}$. C_{SL} and D_{SL} are coupling coefficients at the liquid–solid interface for the repulsive and attractive terms in the Lennard-Jones potential. These two are adjustable terms in the simulation, and they increase as the liquid–solid interaction increases. The results clearly show that the contact angle at equilibrium decreases with increasing liquid–solid interaction. This is intuitively expected since contact angle is expected to decrease as wettability increases.

Fig. 3.1 Graphic vector representation of the three surface tensions acting on the sessile droplet

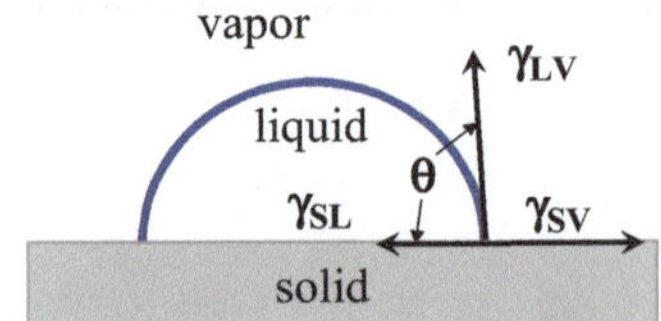

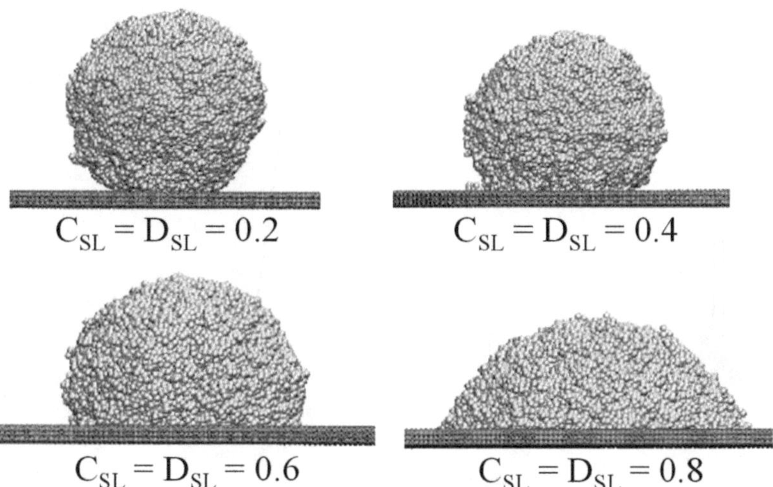

Fig. 3.2 Typical profiles of 40,000 atom drop at equilibrium with varying liquid–solid interaction (simulated with coupling coefficients: $C_{SL}=D_{SL}=0.2, 0.4, 0.6,$ and 0.8) (Reproduced with permission from [4], Copyright 2009 IOP Publishing)

The dynamics of spreading a liquid droplet to its final static state on a solid surface has been modelled by molecular kinetic theory and hydrodynamic theory [5–16]. The molecular kinetic theory emphasizes the importance of liquid–solid interaction during liquid advance. The latter studies the influence of viscosity on the relationship between advancing contact line velocity V and dynamic contact angle θ_D. To date, although a quantitative predictive theory remains elusive, the emerging view is that the advancing contact line will cease to advance when all of its kinetic energy is dissipated. Figure 3.3 depicts a schematic of the liquid wetting and spreading process. Shown in the inset of Fig. 3.3 is a magnified view of the advancing contact line moving at V with a dynamic contact angle θ_D.

When the liquid drop first wets the surface, θ_D is at its theoretical maximum value of 180°. The liquid drop then wets and spreads with an advancing contact line velocity V at dynamic contact angle θ_D. The velocity V and θ_D both decrease as the kinetic energy of the liquid droplet is dissipating due to friction at the liquid–solid interface. When $V=0$, the drop reaches its final position and the static contact angle θ is attained. The velocity for the advancing contact line (V) is given by Eq. (3.2) according to the improved molecular kinetic theory [6, 7].

$$V = \frac{1}{\zeta_0}\gamma_{LV}\left(cos\theta - cos\theta_D\right) \tag{3.2}$$

where γ_{LV} is the surface tension of the liquid, θ is the static contact angle, θ_D is dynamic contact angle at V, and ζ_0 is friction coefficient per unit length of the contact line and is given by:

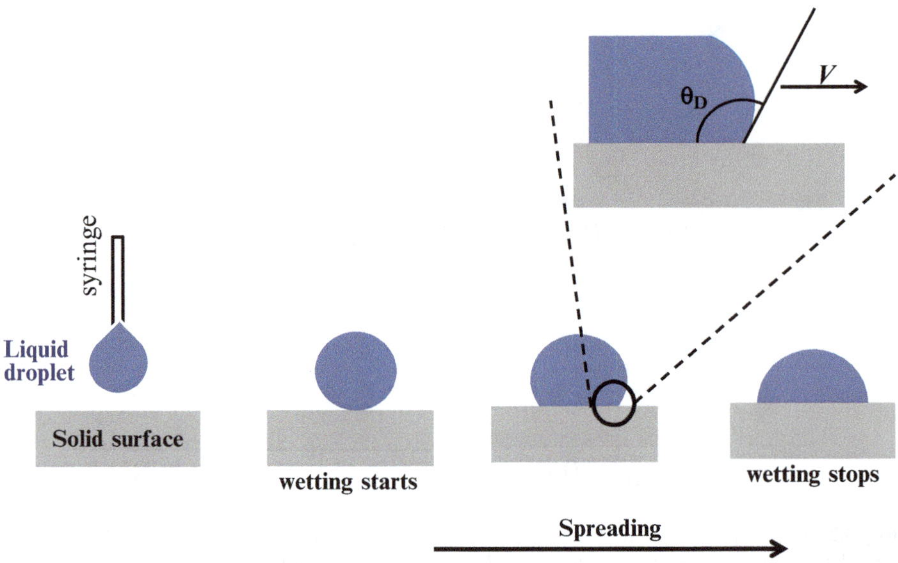

Fig. 3.3 Schematic showing the wetting and spreading of a liquid droplet on a solid surface (Inset: Magnified view of the advancing three-phase zone, V is the velocity of the advancing contact line and $\boldsymbol{\theta}_D$ is the dynamic contact angle which decreases as wetting proceeds)

$$\zeta_0 = \frac{\eta_L \cdot \upsilon_L}{\lambda^3} \exp\left[\frac{W_a}{n \cdot k_B \cdot T}\right] \tag{3.3}$$

η_L is viscosity of the liquid, υ_L is the volume of the unit flow, λ is characteristic length of displacement, W_a is the work of adhesion, n is the number of sites per unit area of solid which the energy is dissipated, k_B is Boltzmann constant, and T is temperature.

It is clear from Eq. (3.2) that the main drivers for a high velocity advancing contact line are strong liquid–solid interaction (low $\boldsymbol{\theta}$) and low friction. The rate of spreading for the liquid droplet can also be modelled using molecular dynamic simulations. Figure 3.4 shows the relaxation of the contact angle ($\boldsymbol{\theta}_D$) as a function of computer time. Three sets of data points are generated for three different liquid–solid interactions, again the stronger the interaction (larger C_{SL} and D_{SL}) the smaller the final contact angle. The three curves shown in the plot are calculations from the molecular kinetic theory. The good fit shown in Fig. 3.4 indicates that the molecular dynamic simulation and the molecular kinetic theory are in agreement in this case.

In reality, wetting and spreading on solid surface is complicated. Typically, the decay of the dynamic contact angle ranges between $t^{-0.1}$ and t^{-1} over time (t) depending on the modes of energy dissipation for the spreading liquid [6, 7, 15]. The mode of energy dissipation during spreading can be dominant by the hydrodynamic of the liquid, where viscosity will be the main driver. It can be dominated by the liquid–solid interaction, which can originate from molecular adhesion for smooth surface or pinning in case of rough surface. Of course, mixed mode is always a possibility. In any event, both modes of energy dissipation have opposite

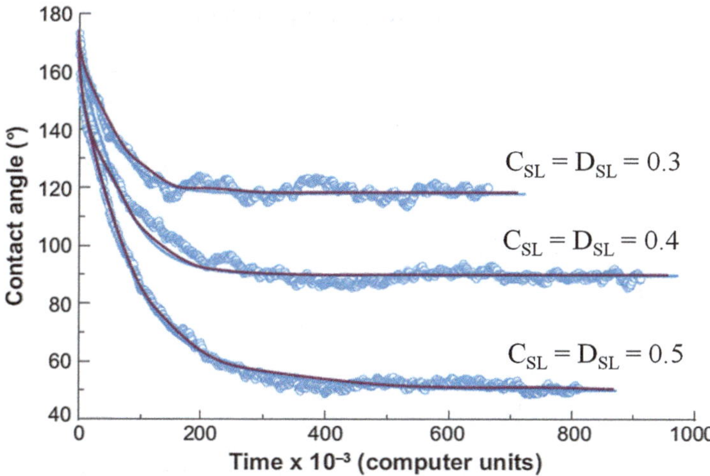

Fig. 3.4 Molecular dynamic simulations for the relaxation of the contact angle as a function of liquid–solid interaction (Reproduced with permission from [8], Copyright 1999 The American Chemical Society)

consequences. For example, a favorable liquid–solid interaction would facilitate liquid advancing (wetting). On the other hand, the strong interaction may increase the frictional force and slow down the advance of the contact line. An illustrative example of the complexity can be found in the work of Vega and co-workers [15] who reported a study on the wetting dynamics of polyethylene terephthalate (PET) fiber by silicone oils in 2005. The kinematic viscosities of the silicone oils (PDMS5, 20, 50, and 500) vary from 5 to 20 to 50 to 500 mm^2 s^{-1} as the molecular weight of PDMS (polydimethylsiloxane) increases. The time-dependent meniscus profile due to wetting was captured by a high-resolution, high-speed camera. Figure 3.5 plots the dynamic contact angles versus time for the four silicone oils. The overall result indicates that there is a strong viscosity effect on the wetting dynamics. Further modelling of the system revealed that there is a difference in wetting dynamics between the low and high viscosity fluid. In the low viscosity regime, wetting is controlled by the interaction between the liquid and the solid. The dynamic follows the molecular kinetic theory and θ_D decays as t^{-1}. As the viscosity is increased, there is a cross-over in regime. The wetting dynamic becomes hydrodynamic control. For example, the decay kinetic for PDMS500 is dominated by viscosity and θ_D decays as t$^{-0.5}$. It is important to note that the surface tensions for these oils are practically the same, 20.5±1 mN/m. The critical surface tension for PET is known to be ~43 mN/m [2], suggesting that all four oils should wet the PET fiber favorably and equally. Based on energetic consideration, the final static contact angles for these oils should all be close to 0°. On the other hand, the data in Fig. 3.5 suggest that, although they all exhibit a low contact angle, their final contact angles are not the same. They decrease in the following order: PDMS500>PDMS50>PDMS20~PDMS5, suggesting that the four silicone oils may have been ending up in four different wetting states on the PET fiber surface.

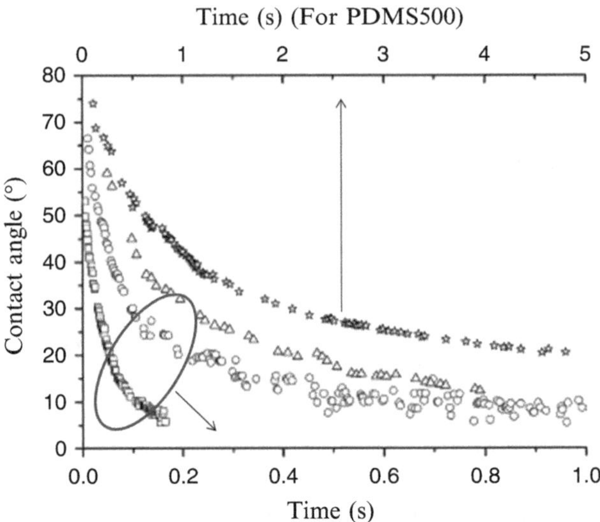

Fig. 3.5 Dynamic contact angle versus time for PDMS5 (*open square*), PDMS20 (*open circle*), PDMS50 (*open triangle*), and PDMS500 (*star*). (Reproduced with permission from [15], Copyright 2005 The American Chemical Society)

3.3   The Four Measurable Contact Angles

One of the faulty perceptions in the surface literature is the constant citation that the droplet for the Young's angle is from a thermodynamically equilibrium state. Young did use the term equilibrium in his paper. He wrote the following sentences related to the state of the sessile droplet on a surface. "*We may therefore inquire into the conditions of equilibrium of the three forces acting on the angular particles, one in the direction of the surface of the fluid only, a second in that of the common surface of the solid and fluid, and the third in that of the exposed surface of the solid.*" Given the era the paper was written, which is 200 years ago, 50 years after Newton developed the classical mechanics, and ~50 years before thermodynamics, what Young clearly meant is that the angle of contact is a result of a mechanical equilibrium at the three phase contact line, not thermodynamic equilibrium. He discussed lengthy about the balance of the wetting force and the cohesion of the liquid in his sessile drop and capillary tube/plate experiments. The persistent misquotation by researchers may be due to their strong bias in thermodynamics. Kinetic factor very often has become a secondary cause or even regarded as exception. What if there is insufficient kinetic energy to bring the drop to the most stable state and the drop is pinned in a metastable state?

It has been known in the literature for a long time that there are three measurable contact angles, θ, θ_A, and θ_R for a given liquid–solid system [17–19]. The static contact angle θ, often called the Young's equilibrium contact angle, lies between the advancing angle θ_A and receding angle θ_R. There were reports as early as 1940 that the Young's angle is not from an equilibrium wetting state. For instances, Fowkes and Harkins [20] reported shaking the liquid–solid interface prior to contact angle measurement, which could bring the system in equilibrium. The resulting contact angles were reported to be reproducible with zero hysteresis. Pease [19] showed that liquid droplets could be converted to a stable wetting state upon slight jarring or after vibration. In recent years, two types of apparatuses have been used to investigate the metastable states between θ_A and θ_R. The first configuration involves immersing a plate (the solid surface) into a liquid and a mechanical noise is supplied to perturb the liquid–solid interface. The changes in the three phase contact line before and after vibration can be monitored as contact angle via the meniscus height or as a wetting force measured by the microbalance in a modified Wilhelmy plate setup [21]. The second lab configuration involves attaching a vibration or noise source to the solid surface. After the sessile drop is dispensed onto the surface, noise of varying amount of energy can be supplied to the sessile droplet. The contact angles before and after vibration are recorded [22]. Representative lab setups are shown in Fig. 3.6.

In 1978, Smith and Lindberg [23] first explored the use of acoustic energy from a loud speaker to influence the water contact angles on different surfaces. They showed that the contact angle θ of water decreases as the input energy increases. The result implies that there are many metastable states between the advancing and receding angles. Andrien and co-workers [24] observed de-pinning of the contact

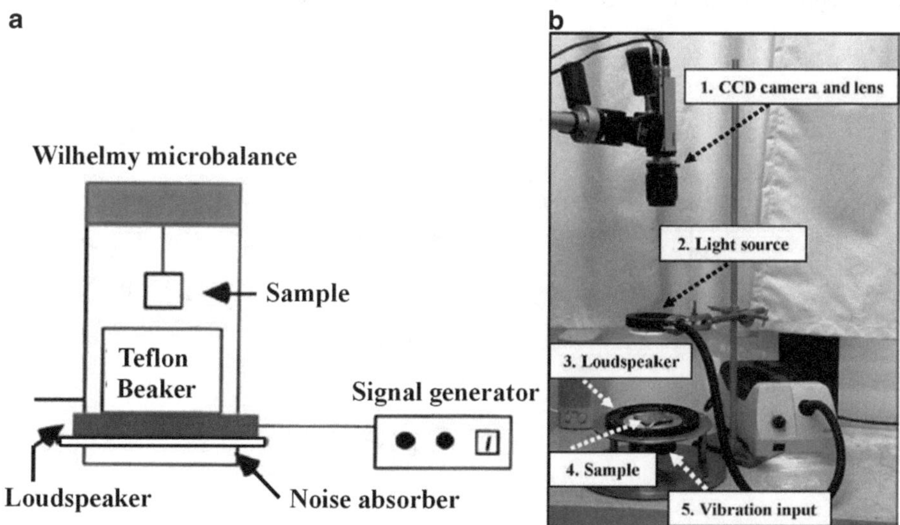

Fig. 3.6 Experimental setups for vibrating (**a**) the contact line and (**b**) the sessile drop (a. Reproduced with permission from [21], Copyright 2002 Elsevier; b. Reproduced with permission from [22], Copyright 2004 Elsevier)

line when a vertical vibration (50 Hz at variable amplitudes) is applied to the water and diiodomethane droplets on various surfaces. Decker and Gareff [25] reported probing the energy barrier between the wetting states of the advancing and receding angle using vibration. By studying the capillary rise of the liquid, they found that advancing angle decreases and receding angle increases as the input energy increase. The implication of this work will be discussed in more detail in the next section. Similarly, Della Volpe and co-workers [21] studied the change in advancing and receding contact angles as a function of vibration energy using a microbalance in a Wilhelmy plate setup (Fig. 3.6a). Similar to Decker and Gareff, they also found that hysteresis decreases as the input energy increases. Both groups also found a good cosine relationship between the equilibrium contact angle (θ_{eq}) and θ_A and θ_R Eq. (3.4).

$$\cos \theta_{eq} = \left(0.5 \cos \theta_A + 0.5 \cos \theta_R\right) \tag{3.4}$$

More recently, Marmur and co-workers [22, 26] reported studies of the effect of vibration on the wetting states of sessile droplets on both rough and smooth surfaces. The results on rough surface will be detailed in Chap. 4. In their 2010 study, they measured the water static contact angle θ as well as the advancing and receding angle (θ_A and θ_R) on 42 smooth surfaces. Vibration of each droplet for 15 s leads to the formation of the most stable droplet with an equilibrium contact angle of θ_{eq}. Their results confirm that there exist four measurable contact angles, namely θ, θ_{eq}, θ_A, and θ_R when a liquid wets a surface. A schematic for these four contact angles is depicted in Fig. 3.7. Figure 3.8 compares the position of θ_{eq} relative to the other three contact angles. In agreement with the literature, the static contact angle θ lies between θ_A and θ_R. θ_{eq} is also between θ_A and θ_R, but it is different from θ. In a slight disagreement with early work, Marmur and co-workers [26] showed that θ_{eq} is not

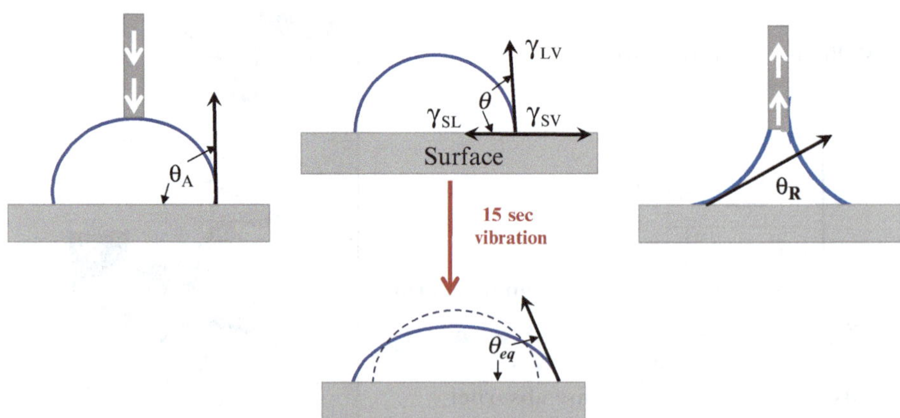

Fig. 3.7 Schematic of the four measurable contact angles for a given sessile droplet

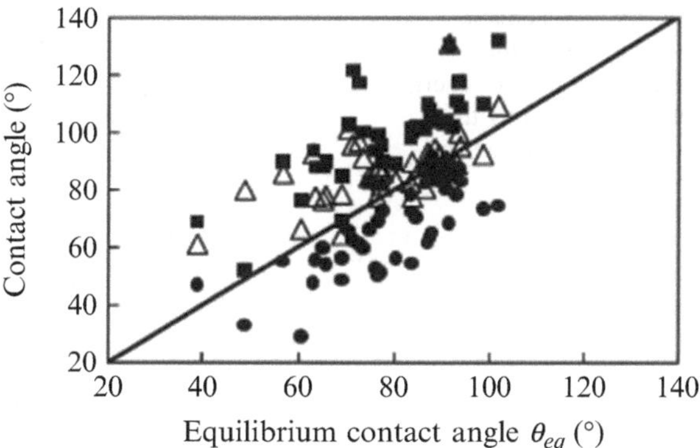

Fig. 3.8 Comparison among the advancing (*filled square*), receding (*filled circle*) and static contact angles (*open triangle*) with the equilibrium contact angle (*solid line*) (Reproduced with permission from [26], Copyright 2010 The American Chemical Society)

equal to the average of θ_A and θ_R (or the cosine average according to Eq. (3.4)) among the 42 smooth surfaces. In any event, the significance of the 2010 study is its clear demonstration that *static contact angles determined on smooth surfaces are all from the metastable wetting states*.

3.4   Wetting States on Smooth Surfaces

The thermodynamic of the wetting process has been modelled by Johnson and Dettre [27], Neumann and Good [28] and more recently by Long and co-workers [29]. Conceptually, wetting of a surface involves moving the advancing contact line as the liquid spreads. According to both molecular kinetic theory and hydrodynamic theory, the advancing contact line will cease to move when all of its kinetic energy is dissipated by friction. This friction can be from roughness or adhesion interaction between the liquid and the surface. Theoretically, the friction can be viewed as a series of energy barrier the advancing liquid has to overcome. A schematic showing the changes in Gibbs free energy and contact angle as wetting proceeds is given in Fig. 3.9. The process of wetting simply involves the liquid traveling through a series of energy barrier. When the liquid first wets the surface, the dynamic contact angle is essentially 180°. As it spreads, the kinetic energy is dissipating due to frictions created by the energy barriers. When all of the kinetic energy is consumed, the liquid droplet will be stuck in a metastable wetting state. If sufficient kinetic energy is provided to the liquid droplet, such as through vibration, the contact line will be de-pinned and the liquid droplet will force to spread and overcome all the energy barriers. It will end up in the thermodynamically stable wetting state with a contact angle of θ_{eq}. The wetting states due to liquid advancing and receding are always metastable, and their respective angles are known to be the largest and smallest among the four measurable angles. The relative position of the four measurable contact angles: θ , θ_{eq}, and θ_A and θ_R are shown in Fig. 3.9.

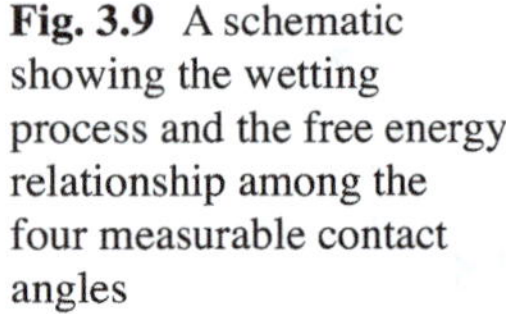

Fig. 3.9 A schematic showing the wetting process and the free energy relationship among the four measurable contact angles

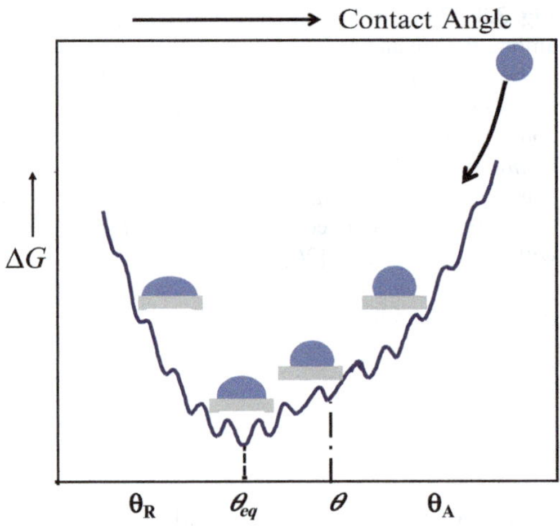

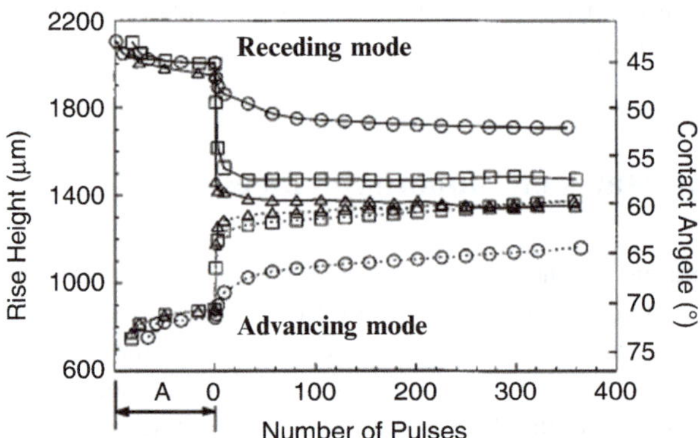

Fig. 3.10 Relaxation of capillary rise heights and contact angles as a function of varying amount of inputting vibration energy: A 3 min ambient relaxation before vibration (*open circle*) 250 μm upward pulse, (*open square*) 500 μm upward pulse, (*open triangle*) 1000 μm upward pulse (Reproduced with permission from [25], Copyright 1996 The American Chemical Society)

Evidence for the existence of multiple wetting states between the advancing and receding metastable states comes from the study of Decker and Gareff [25]. These authors studied the advancing and receding contact angles through a capillary rise experiment. The advancing or receding contact angles are calculated from the height of the meniscus when the glass substrate is immersing or withdrawing from a beaker of water in a Teflon beaker. Variable amount of vibration energy was supplied to the wetting contact lines, and the results are summarized in Fig. 3.10. There are two important messages from these data. First, with the maximum amount of energy supplied via a lot of pulses, the contact line reaches the most stable state and θ_{eq} is obtained. Secondly, the results clearly show that different metastable state can be populated depending on the amount of energy supplying to the system, in both advancing and receding modes. As the energy supplied to the contact line is increased, both advancing and receding metastable states would continue to drive toward the equilibrium state. Very similar results were obtained by Mettu and Chaudhury [30]. They reported the use of white noise vibration to drive the water droplets from the "receding" and "advancing" states to the most stable equilibrium state, on PDMS surface. In the low noise regime, they were able to populate intermediate advancing and receding states analogously to those reported by Decker and Gareff.

3.5   What Determines Contact Angle? Contact Line or Contact Area

As described earlier, the Young's equation is merely a mechanical description of the balance of the three forces that act on the three phase contact line. The concept and the measurement of the contact angle are deceptively simple, but the Young's equation has been a source of argument for a long time. Two of the four quantities, γ_{SV} and γ_{SL}, in the equation cannot be measured reliably. These two quantities are related to the surface tension of the solid and the interfacial tension between the liquid and the solid. They are not only of academic interest, but also relevant to materials design and many industrial applications. According to Zisman [2], earlier researchers tried to solve the Young's equation using thermodynamics. Dupre [31] introduced the concept of reversible work of adhesion (W_A) between the liquid and the solid surface and its relation to γ_{SV} and γ_{SL} is given by:

$$W_A = \gamma_{SV} + \gamma_{LV} - \gamma_{SL} \tag{3.5}$$

This equation is basically a thermodynamic expression that the work done to separate the liquid droplet from the solid surface must be equal to the free energy changes of the system. In other words, the thermodynamic approach would suggest that the interactive energy in the contact area beneath the liquid droplet is critically important in determining the contact angle. A 3D representation showing partial wetting of the solid surface by the liquid droplet is shown in Fig. 3.11.

The validity of the thermodynamic approach has been a controversial subject. Conceptually, W_A is the free energy of adhesion between the liquid and the surface divided by the wetted area. Its unit is mN/m. The unit for the three surface tensions in the right hand side of Eq. (3.5) is also mN/m. Although the unit between them is the same, they are different fundamentally. In 1965, Gray [32] wrote: "*It is clear from these definitions that surface tension and surface-free energy are quite distinct quantities. Surface tension is a tensor, which act perpendicularly to a line in a surface. It is the quantity involved in contact angle equilibrium. Surface-free energy is a scalar quantity without directional properties and it is a property of an area of the surface. It is the quantity involved in thermodynamic properties of surfaces.*"

Pease [19] was the first to point out that the air–liquid interface in contact with the solid is a one-dimensional system (the contact line), not a two-dimensional

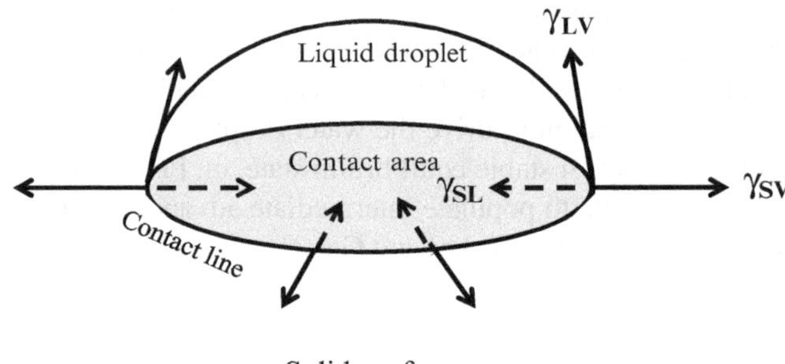

Fig. 3.11 A 3D view showing partial wetting of a solid surface by a liquid droplet

system (contact area) (Fig. 3.11). Any work of adhesion calculated from the contact angle would only reflect on work done on the contact line. This is different from the thermodynamic approach, which optimizes the free energy of the entire wetted area beneath the drop. He concluded that *"the application of Young's equation to equilibrium angle is not at all a measure of the overall mean work of adhesion between solid and liquid."* Bartell and Shepard [33] also concluded that contact angle is determined by the contact line at the liquid–solid–air interface, not the area beneath the drop. In their study of the wetting of paraffin surfaces with controlled roughness, they found that contact angle is insensitive to the height of the asperities (which lead to increased surface area), but sensitive to the inclination angle of the asperities. In 1959, Johnson [34] pointed out the lack of rigor by surface researchers in his era when dealing with the quantities in the Young's equation. Terms such as surface tension, surface energy, and surface-free energy were used as desired. This had contributed to the mass confusion in the literature. He went on to derive the Young's equation using the technique of Gibbs. Gravity and adsorption were explicitly considered. He found that Young's equation is a relationship among different surface tensions, not surface-free energies.

While progress has certainly been made in understanding the physics and thermodynamics of the wetting process, the state of confusion in the surface literature on the other hand is continuing. It appears that this situation can only be improved if the theory is back up by experimental facts. In 2003, Extrand [35] reported the fabrication of two smooth surfaces (verified by AFM) comprising chemically heterogeneous domains several mm in size in the center of the sample. The configuration of the surfaces studied is shown in Fig. 3.12a. The first heterogeneous surface was prepared by coating a thin layer of polystyrene from a dilute solution onto an oxygen plasma-cleaned silicon wafer. The polystyrene domain is hydrophobic, whereas the silicon wafer is hydrophilic. The second surface was prepared by etching a small area in a PFA (perfluoroalkoxy) film with a sodium naphthalene complex solution. In this case, the etched area is hydrophilic, whereas the rest of the surface is hydrophobic. The advancing contact angle data for the materials in these two surfaces are summarized in Table 3.1. The changes in contact angles during the advancing contact angle experiments with Surfaces **1** and **2** are given in Fig. 3.12b and the contact angle data are included in Table 3.1. Initially, a small water droplet was gently

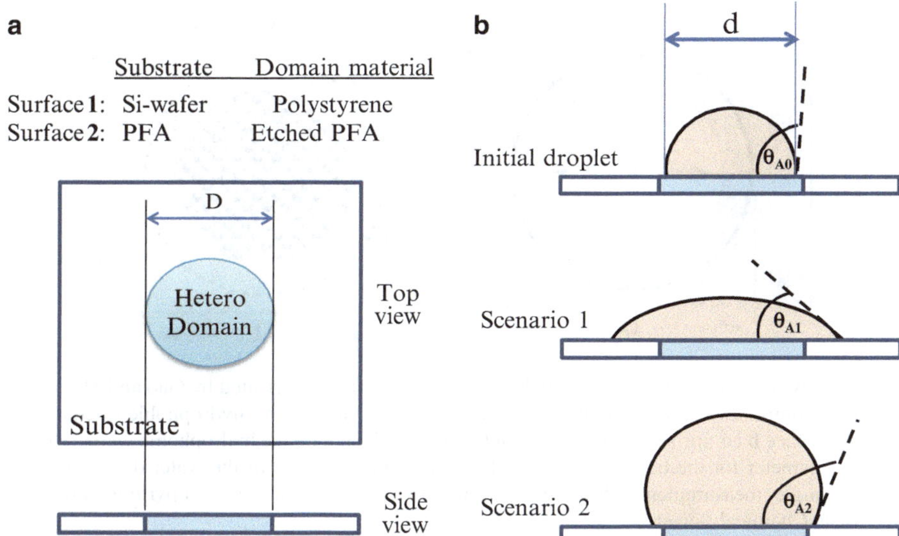

Fig. 3.12 (**a**) Configuration and materials for the two chemically heterogeneous surfaces **1** and **2**, (**b**) water contact angles at $t=0$ and after the contact lines advance across the material boundary

Table 3.1 Contact angle data for materials and surfaces **1** and **2** (data taken from [35])

	θ_A	θ_{A0}	θ_{A1}	θ_{A2}
Si-wafer	~7°			
Polystyrene	95°			
Etched PFA	67°			
PFA	109°			
Surface **1**		95°	~6°	
Surface **2**		67°		109°

deposited in the center of the polystyrene domain in Surface **1** ($d<D$). The $t=0$ θ_{A0} was measured at 95°. As small amount of water is added to the droplet, its volume expands. When the contact line advances beyond the polystyrene domain and makes contact with the silicon wafer ($d\geq D$), a drastic decrease in contact angle is observed (Scenario 1). The measured θ_{A1} is ~6°, identical to that of the oxygen plasma-treated silicon wafer. As for Surface **2**, θ_{A0} was measured at 67°, same as the etched PFA control. As the contact line advance across the etched PFA domain, there is an increase in contact angle (Scenario 2) and θ_{A2} was measured at 109°, identical to that of the PFA control. In these two experiments, at the moment the advance contact line crosses the polystyrene or etched PFA domains, the change in surface-free energies is small. In fact, the average surface-free energy in the contact area can be estimated based on the solid area fraction of the materials involved. *In any event, the fact that θ_{A1} and θ_{A2} is dictated by the materials at the three phase contact line rather than the materials in the contact area positively proof that it is the contact line not the contact area determines the magnitude of the contact angle.*

A very similar experimental design was reported later by Gao and McCarthy, whose objective was to use the experimental data to argue "How Wenzel and Cassie

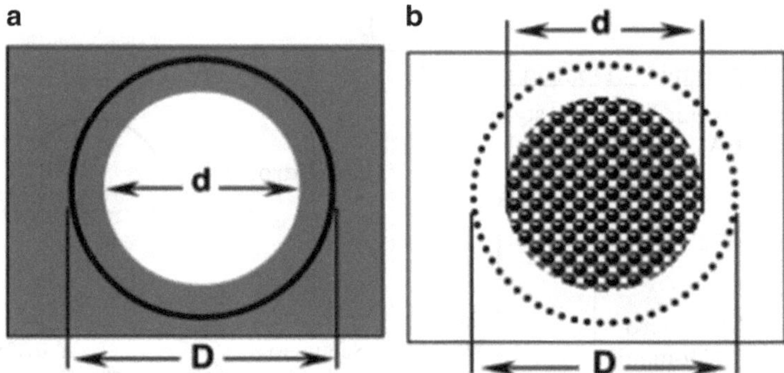

Fig. 3.13 Schematics of the chemically heterogeneous surfaces fabricated by Gao and McCarthy, Surface **a** comprises a partially hydrolyzed hydrophilic spot (*white*) on a hydrophobic silicon wafer (*gray*); Surface **b** comprises a superhydrophobic (textured) spot on the hydrophobic wafer (*white*); d is the diameter for the heterogeneous spot; and *D* is the diameter of the water droplet used in contact angle measurements (Reproduced with permission from [36], Copyright 2007 The American Chemical Society)

Were Wrong?" [36]. More detailed discussion on the story of Wenzel and Cassie will be given in Chap. 4. Here, the two surface designs (**a** and **b**) in Gao and McCarthy's work are depicted in Fig. 3.13. Sample **a** comprises a hydrophilic spot within a hydrophobic silicon surface. The sample was prepared by first making a cleaned silicon wafer hydrophobic with a perfluoroalkyldimethylchlorosilane, followed by etching a spot with a sodium hydroxide solution. The advancing/receding contact angles (θ_A/θ_R) for the hydrophobic and hydrophilic areas are found to be (119°/110°) and (35°/10°), respectively. Experimentally, three different spot diameters ($d = 1$, 1.5, and 2 mm) with varying water droplet diameters (D) were studied. The (θ_A/θ_R) results are summarized in Table 3.2. Without exception, when the diameter of the water droplet is smaller than the diameter of the hydrophilic spot ($d > D$), the contact angle data indicate that the drop is in the hydrophilic area. On the other hand, when $d < D$, the contact angle data indicate that the surface is hydrophobic. This is despite the fact that majority of the contact area is hydrophilic.

Similar results were obtained with surface **b**. Surface **b** was prepared by first texturing a rough spot of diameter d on a cleaned silicon wafer followed by a fluorosilane treatment. The θ_A/θ_R ratios for the textured and smooth areas are found to be 168°/132° and ~117°/~82°, respectively. The textured area is superhydrophobic, and the smooth area is hydrophobic. Again, three different spot diameters ($d = 1$, 1.5, and 2 mm) with varying water droplet diameters (D) were studied and the results are also included in Table 3.2. Similar to the results obtained with Surface **a**, the contact angle data with Surface **b** show that the droplet reveals superhydrophobicity when $d > D$ and hydrophobicity when $d < D$. Again, the size of the contact angle is dictated by the location of the contact line, not the surface energetic of the contact area.

In summary, the experimental data from Extrand as well as Gao and McCarthy leave little doubt that the energetics at the three phase contact line determines the size of the contact angle, not the contact area. Additional support for this conclusion comes from Bormashenko [37], who reported a detailed thermodynamic analysis of

Table 3.2 Water contact angle data for Surfaces **a** and **b** as a function of spot size and water droplet diameter (data taken from [36])

		d (mm)	D (mm)	θ_A/θ_R
Surface **a**		1	0.5	33°/11°
		1	1.5	119°/110°
		1	2.0	118°/108°
		1	2.5	119°/108°
		1.5	0.7	35°/9°
		1.5	2.0	120°/110°
		1.5	2.5	118°/109°
		1.5	3.0	120°/111°
		2	0.7	35°/10°
		2	2.5	120°/110°
		2	3.0	119°/110°
		2	3.5	118°/111°
Surface **b**		1	0.5	168°/32°
		1	1.1	117°/81°
		1	1.2	117°/82°
		1	1.3	117°/81°
		1.5	0.7	166°/134°
		1.5	1.6	117°/82°
		1.5	1.7	117°/81°
		1.5	1.8	117°/82°
		2	0.7	165°/133°
		2	2.1	117°/82°
		2	2.2	117°/81°
		2	2.3	118°/82°

the wetting of smooth as well as composite surfaces and concluded that contact angle is a one-dimensional, not two-dimensional phenomenon.

3.6   Effects of Solvent and Temperature

From Eq. (3.1) to Eq. (3.3), surface tension, viscosity, and ambient temperature are parameters that will have an effect on the measured contact angle. While viscosity will primarily influence the rate of relaxation of θ_D during wetting, surface tension will have effects on both the relaxation rate Eq. (3.2) and the final contact angle Eq. (3.1). Surface tension is a property of liquid caused by uneven attraction of liquid molecules (or surface cohesion) at or near the surface, the higher the surface tension, the larger the resistance for the liquid to wet. The effect of solvent on contact angle has been studied on many polymer surfaces, such as poly(tetrafluoroethylene), poly(vinylidene fluoride), poly(ethylene), poly(vinyl chloride), poly(vinylidene chloride), polystyrene, and the like and adsorbed monolayers from aliphatic fatty acids and perfluoro fatty acids. The subject was summarized well by Zisman [2]. The solvent effect is usually plotted as cos θ versus γ_{LV} of the liquid. A typical plot

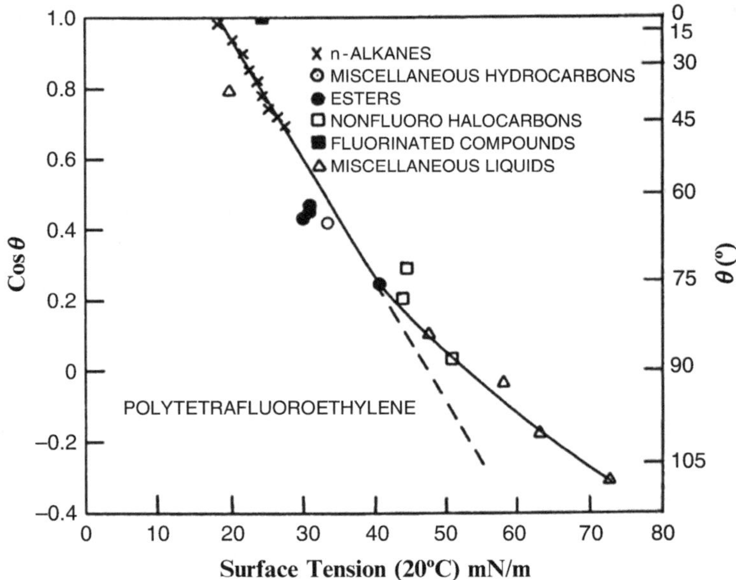

Fig. 3.14 Plot of cos θ versus surface tension of the liquid for poly(tetrefluoroethylene) (Reproduced with permission from [2], Copyright 1964 The American Chemical Society)

showing the data taken on poly(tetrafluoroethylene) is given in Fig. 3.14. cos θ versus γ_{LV} exhibits a linear relationship in the low γ_{LV} regime. At $\gamma_{LV} > 40$ mN / m, the relationship becomes nonlinear and is attributable to H-bonding and polar interactions between the liquid and the solid surface. The critical surface tension, γ_C, is defined by the intercept at cos $\theta = 1.0$. As it turns out, γ_C was found to be the most valuable parameter in the plot because it is the characteristic of the surface tension of the solid surface [2]. The plot, known as Zisman plot today, is still a useful methodology in gaining information about the surface energetics of new surface materials [38].

The effect of temperature on contact angle has seldom been studied. However, as the printing industry is evolved, printing has become a manufacturing technology for additive manufacturing and printed electronics. Inks are no longer just printed on paper at room temperature, they have been printed on different substrates, including plastics and metal foils under a variety of conditions, e.g., temperature higher than ambient. Fundamental understanding of how liquid interacts with solid surface has shown to be crucial not only to the functional performance of the printed device [39, 40] but also to its resolution [41]. Mettu, Kanungo, and Law [42] recently reported a wetting study of a UV ink monomer (neopentyl glycol diacrylate, SR-9003) on four different coated substrates (DTC-coated and Flexo-coated biaxial oriented polypropylene (BOPP) and SGE paper) at temperatures ranging from ~22° to 95 °C. These substrates are macroscopically smooth, but do comprise micro/nano scale roughness. Figure 3.15 summarizes the results on the time-dependent wetting behavior of the UV ink monomer on these four substrates at ~22° and 95 °C.

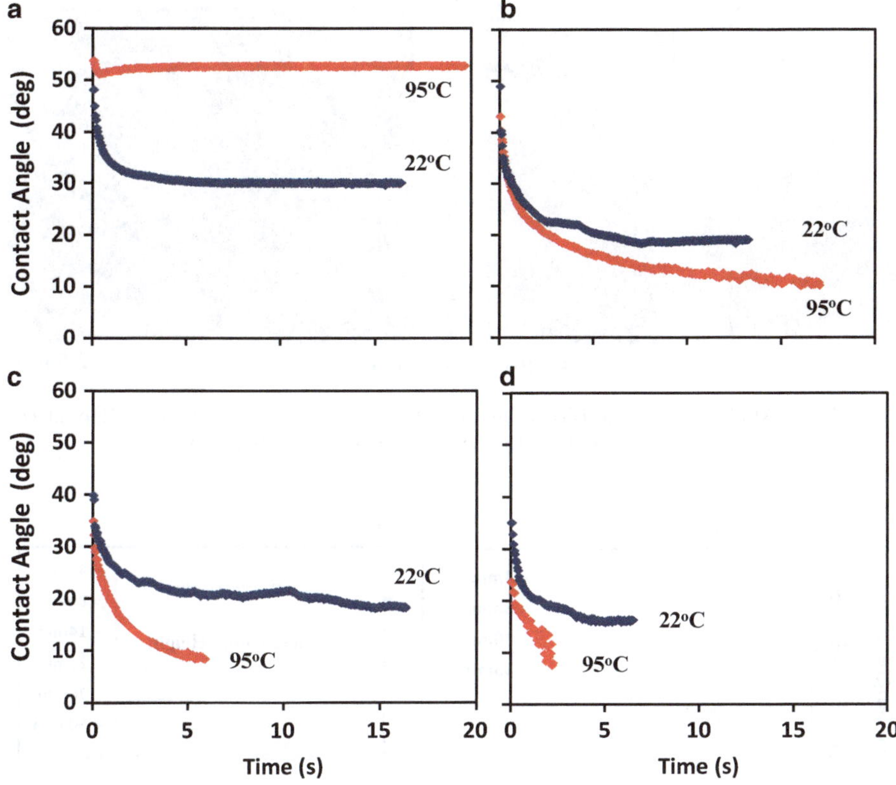

Fig. 3.15 Plots of the time-dependent contact angles for the monomer at substrate temperatures of 22 °C and 95 °C on (**a**) DTC-coated BOPP, (**b**) Flexo-coated BOPP, (**c**) DTC-coated SGE, and (**d**) Flexo-coated SGE (Reproduced with permission from [42], Copyright 2013 The American Chemical Society)

The results in Fig. 3.15b–d indicate that increase of temperature results in an increase of the rate of relaxation for the contact angle and a decrease of the final contact angle. Since both surface tension and viscosity are known to decrease as temperature increases, the increase in relaxation rate is certainly rational. The decrease in the final contact angle can attribute to the temperature effect on surface tension. On the other hand, an anomalous temperature effect is observed on the DTC-coated BOPP substrate (Fig. 3.15a). The final contact angle at 95 °C is higher than that at 22 °C (~52° vs. 30°). Study of the effect of temperature on the surface morphology by AFM shows that nano protrusions were developed as the polypropylene fibers in BOPP undergo thermal expansion underneath the DTC coating (Fig. 3.16). The nano protrusions create friction against liquid spreading, resulting in pinning of the liquid droplet as the liquid contacts the rough substrate surface at 95 °C. This interpretation is supported by measuring the contact angle obtained from different dispensing drop heights. The results are summarized in Fig. 3.17a, b. At 22 °C, the

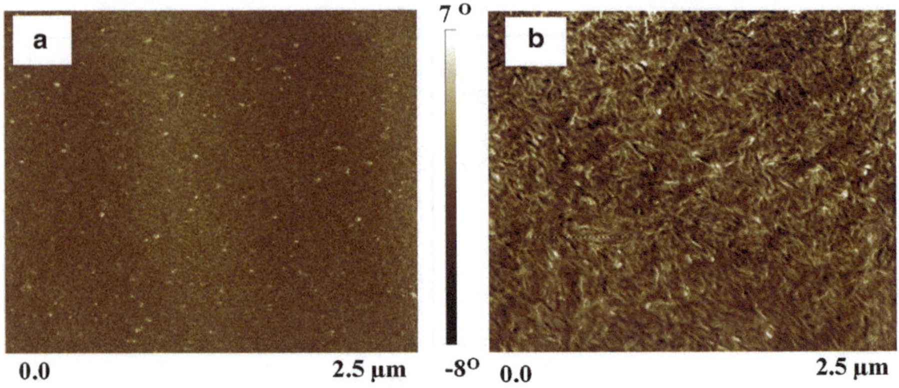

Fig. 3.16 AFM phase images of DTC-coated BOPP (**a**) at 22 °C and (**b**) at 95 °C (Reproduced with permission from [42], Copyright 2013 The American Chemical Society)

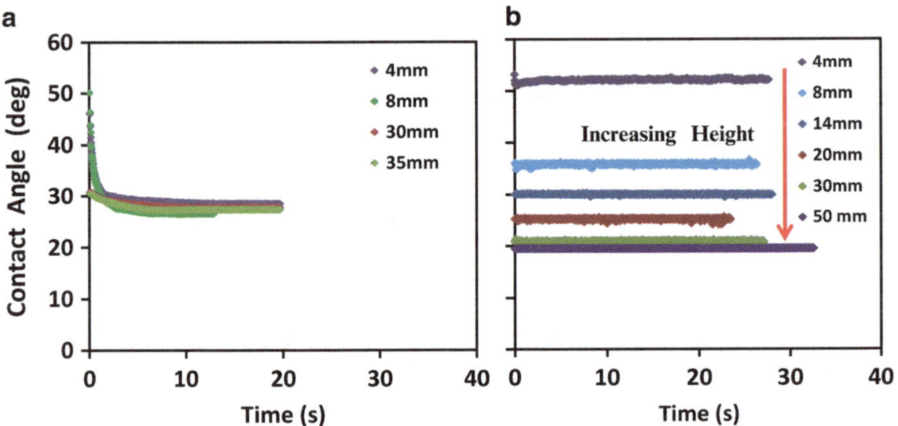

Fig. 3.17 Plots of time-dependent contact angles for the monomer on DTC-coated BOPP substrate at varying drop height at (**a**) 22 °C and (**b**) 95 °C (Reproduced with permission from [42], Copyright 2013 The American Chemical Society)

final contact angle is at ~30° and is insensitive to the drop height as it varies between 4 and 35 mm. On the other hand, the final contact angle is shown to be sensitive to the drop height at 95 °C, the higher the drop height the smaller the final contact angle. The result is attributable to the additional kinetic energy provided to the wetting droplet, consistent with the wetting model shown in Fig. 3.9.

This work demonstrates that care should be taken in interpreting the wetting data at different temperature. There are two components in liquid–solid interactions. While temperature certainly has effects on the surface tension and viscosity of the liquid, it may also have an effect on the morphology of the solid surface too.

References

1. Young T (1805) An essay on the cohesion of fluids. Phil Trans R Soc London 95:65–87
2. Zisman WA (1964) Relation of the equilibrium contact angle to liquid and solid constitution. In: Fowkes F (ed) Contact angle, wettability, and adhesion, advances in chemistry. American Chemical Society, Washington, DC, pp 1–51
3. Shuttleworth R, Bailey GJ (1948) The spreading of a liquid over a rough surface. Discuss Faraday Soc 3:16–22
4. Bertrand E, Blake TD, De Coninck J (2009) Influence of solid-liquid interactions on dynamic wetting: a molecular dynamics study. J Phys Condens Matter 21:464124
5. Duvivier D, Blake TD, De Coninck J (2013) Towards a predictive theory of wetting dynamic. Langmuir 29:10132–10140
6. De Coninck J, Blake TD (2008) Wetting and molecular dynamics simulations of simple liquids. Annu Rev Mater Res 38:1–22
7. Seveno D, Vaillant A, Rioboo R, Adao H, Conti J, De Coninck J (2009) Dynamics of wetting revisited. Langmuir 25:13034–13044
8. Ruijter MJ, Blake TD, De Coninck J (1999) Dynamic wetting studies by molecular modeling simulations of droplet spreading. Langmuir 15:7836–7847
9. Bonn D, Eggers J, Indekeu J, Meunier J, Rolley E (2009) Wetting and spreading. Rev Modern Phys 81:739–805
10. Ray S, Sedev R, Priest C, Ralston J (2008) Influence of the work of adhesion on the dynamic wetting of chemically heterogeneous surfaces. Langmuir 24:13007–13012
11. Vega MJ, Gouttiere C, Seveno D, Blake TD, Voue M, De Coninck J, Clark A (2007) Experimental investigation of the link between static and dynamic wetting by forced wetting of nylon filament. Langmuir 23:10628–10634
12. Blake TD, De Coninck J (2002) The influence of solid-liquid interactions on dynamic wetting. Adv Colloid Interface Sci 96:21–36
13. Voinov OV (1976) Hydrodynamic of wetting. Fluid Dyn 11:714–721
14. Cox RG (1986) The dynamics of the spreading of liquids on a solid surface. Fluid Mech 16:169–194
15. Vega MJ, Seveno D, Lemaur G, Adao MH, De Coninck J (2005) Dynamics of the rise around a fiber: experimental evidence of the existence of several time scales. Langmuir 21:9584–9590
16. Ranabothu SR, Karnezis C, Dai LL (2005) Dynamic wetting: hydrodynamic or molecular kinetic? J Colloid Interface Sci 288:213–221
17. Bartell FE, Wooley AD (1933) Solid-liquid-air contact angles and their dependence upon the surface condition of the solid. J Am Chem Soc 55:3518–3527
18. Macdougall G, Ockrent C (1942) Surface energy relations in liquid/solid systems. 1. The adhesion of liquids to solids and a new method of determining the surface tension of liquids. Proc R Soc Lond A 180:151–173
19. Pease DC (1945) The significance of the contact angle in relation to the solid surface. J Phys Chem 49:107–110
20. Fowkes FM, Harkins WD (1940) The state of monolayers adsorbed at the interface solid-aqueous solution. J Am Chem Soc 62:3377–3386
21. Della Volpe C, Maniglio D, Morra M, Siboni S (2002) The determination of a 'stable-equilibrium' contact angle on a heterogeneous and rough surfaces. Colloids Surf A Physicochem Eng Asp 206:47–67
22. Meiron TS, Marmur A, Saguy IS (2004) Contact angle measurement on rough surfaces. J Colloid Interface Sci 274:637–644
23. Smith T, Lindberg G (1978) Effect of acoustic energy on contact angle measurements. J Colloid Interface Sci 66:363–366
24. Andrien C, Sykes C, Brochard F (1994) Average spreading parameter on heterogeneous surfaces. Langmuir 10:2077–2080

25. Decker EL, Gareff S (1996) Using vibrational noise to probe energy barrier producing contact angle hysteresis. Langmuir 12:2100–2110
26. Cwikel D, Zhao Q, Liu C, Su X, Marmur A (2010) Comparing Contact angle measurements and surface tension assessments of solid surfaces. Langmuir 26:15289–15294
27. Johnson RG, Dettre RH (1964) Contact angle hysteresis III. Study of an idealized heterogeneous surface. J Phys Chem 68:1740–1750
28. Neumann AW, Good RJ (1972) Thermodynamics of contact angles 1. Heterogeneous solid surface. J Colloid Interface Sci 38:341–358
29. Long J, Hyder MN, Huang RYM, Chen P (2005) Thermodynamic modeling of contact angles on rough heterogeneous surfaces. Adv Colloid Interface Sci 118:173–190
30. Mettu S, Chaudhury MK (2010) Stochastic relaxation of the contact line of a water droplet on a solid substrate subjected to white noise vibration: role of hysteresis. Langmuir 26:8131–8140
31. Dupre A (1869) Theorie mechanique de la chaleur. Gauthier-Villars, Pairs, p 369
32. Gray VR (1965) Surface aspects of wetting and adhesion. Chem Ind 23:969–977
33. Bartell FE, Shepard JW (1953) Surface roughness as related to hysteresis of contact angles. II. The systems of paraffin-3 molar calcium chloride solution-air and paraffin-glycerol-air. J Phys Chem 57:455–458
34. Johnson RE (1959) Conflicts between Gibbsian thermodynamics and recent treatments of interfacial energies in solid-liquid-vapor systems. J Phys Chem 63:1655–1658
35. Extrand CW (2003) Contact angles and hysteresis on surfaces with chemical heterogeneous Islands. Langmuir 19:3793–3796
36. Gao L, McCarthy TJ (2007) How Wenzel and Cassie were wrong. Langmuir 23:3762–3765
37. Bormashenko EA (2009) Variation approach of wetting of composite surfaces: is wetting of composite surfaces a one-dimensional or two-dimensional phenomenon? Langmuir 25:10451–10454
38. Chhatre SS, Guardado JO, Moore BM, Haddad TS, Mabry JM, McKinley GH, Cohen RE (2010) Fluoroalkylated silicon-containing surface—estimation of solid surface energy. ACS Appl Mater Interfaces 2:3544–3554
39. Kang B, Lee WH, Cho K (2013) Recent advances in organic transistor printing processes. ACS Appl Mater Interfaces 5:2302–2315
40. Fukuda K, Sekine T, Kumake D, Tokito S (2013) Profile control of inkjet printed silver electrodes and their application to organic transistors. ACS Appl Mater Interfaces 5:3916–3920
41. Arias AC, Mackenzie JD, McCulloch I, Pivnay J, Salleo A (2010) Materials and applications for large area electronics solution-based approaches. Chem Rev 110:3–24
42. Mettu S, Kanungo M, Law KY (2013) Anomalous thermally induced pinning of a liquid drop on a solid substrate. Langmuir 49:10665–10673

Chapter 4
Wetting on Rough Surfaces

Abstract There are two possible wetting states, Wenzel and Cassie-Baxter, when liquid wets a rough surface. In the Wenzel state, liquid fully wets every area of the rough surface. For hydrophilic material, roughness enhances wettability and results in superhydrophilicity. On the other hand, roughness increases the surface's resistance to wet for moderately hydrophilic and hydrophobic material. The advancing contact line sometimes prematurely pins the liquid droplet into a metastable wetting state, resulting in an anomalously large contact angle. Vibration of the drop de-pins the contact line and relocates the droplet to an equilibrium position with a smaller equilibrium contact angle θ_{eq}. The calculated Wenzel angle agrees well with θ_{eq} confirming that vibration leads to the most stable wetting state on the rough surface. Roughness geometry is shown to have a profound effect on the wetting and spreading process. While surface with cavities and pores wets similarly to the smooth surface, bumps on the other hand interact with the contact line, they retard contact line advancing during spread and drag the contact line during receding. In the case of the Cassie–Baxter state, pockets of air are trapped during liquid wetting, forming a liquid–solid–air composite interface. This interface is characterized by a large contact angle along with a small sliding angle. Surface texture/roughness, low surface energy material, and re-entrant geometry are key design parameters for both superhydrophobicity and superoleophobicity. Since the fully wetted Wenzel state is usually more stable, a lot of attention has been paid to stabilize the Cassie–Baxter state by increasing the energy barrier between them. Hierarchical roughness structure and re-entrant angle at the liquid–solid–air interface are shown to be key enablers, not only to stabilize the Cassie–Baxter composite state from transitioning to the Wenzel state, but also to increase its resistance to collapse when an external pressure is applied. Cassie–Baxter composite state can also be formed on groove surfaces, which will lead to directional wetting. Droplets are shown to move faster in the direction parallel to the grooves through wetting of the solid strips. This is evident by imaging the advancing contact line with a hot polyethylene wax. In the orthogonal direction, the contact line advances by hopping from one solid strip to another. This increases the chance of pinning and results in both large contact angle and sliding angle. With appropriate surface texturing, surface with interesting unidirectional spreading ability has been reported. Despite the fascinating wetting properties and its numerous application potentials, technology implementation of rough surfaces is lagged. The major hurdle for crossing the chasm between research and product is discussed.

© Springer International Publishing Switzerland 2016
K.-Y. Law, H. Zhao, *Surface Wetting*, DOI 10.1007/978-3-319-25214-8_4

Keywords Wetting • Rough surfaces • Wenzel equation • Cassie–Baxter equation • Wettability • Wetting dynamics • Roughness factor • Roughness geometry • Drop vibration • Metastable wetting state • Most stable wetting state • Superhydrophilicity • Antifogging • Self-cleaning • Superhydrophobicity • Wenzel Cassie–Baxter transition • Superoleophobicity • Re-entrant angle • Directional wetting • Unidirectional wetting

4.1   The Two Classic Wetting Models

The concept of wetting a rough surface was first described by Wenzel [1]. Using the thermodynamic argument, he stated that if a liquid wets a solid surface favorably, its wettability will be enhanced on the rough surface. Similarly, if the surface resists wetting, its resistance against wetting will increase when the surface becomes rough. The increase in wettability in the former or wetting resistance in the latter is attributed to the increase in surface area as the surface is roughened. The apparent contact angle on a fully wetted rough surface is given by the Wenzel equation Eq. (4.1)

$$\cos \theta_w = r \cdot \cos \theta \tag{4.1}$$

where θ_w is the Wenzel angle, θ is the contact angle of the smooth surface of the same material, and r is the roughness factor. r is given by:

$$r = \text{actual surface area} \,/\, \text{projected surface area} \tag{4.2}$$

In 1944, Cassie and Baxter [2] extended the analysis of the apparent contact angles for the wetting of porous surfaces similar to those encountered in textiles in clothing and feathers in birds. When a liquid wets a porous surface, air pockets are formed and the liquid–surface interface becomes a composite interface. Again, based on a simple thermodynamic argument, the apparent contact angle is determined by the energetics of the contact area under the liquid droplet, which has two components: one governs by the area fraction of the solid and the other by the area fraction of air. The general expression for the apparent contact angle (θ_{app}) is:

$$\cos\theta_{app} = f_1 \cdot \cos\theta_1 + f_2 \cdot \cos\theta_2 \tag{4.3}$$

where f_1 and f_2 are the area fractions and θ_1 and θ_2 are the contact angles for the two components at the liquid–solid–air composite interface, respectively.

Since one of the components (f_2) is air, $\cos 180° = -1$, Eq. (4.3) becomes the famous Cassie–Baxter equation:

$$\cos \theta_{CB} = f \cdot \cos \theta + \left(f - 1 \right) \tag{4.4}$$

where θ_{CB} is the Cassie–Baxter angle, f is the solid-area fraction, and θ is the contact angle of the smooth surface of the same material.

As elucidated in Chap. 3, the angle of contact formed between a solid surface and the wetting liquid is a result of the mechanical equilibrium of the three surface tensions (γ_{SV}, γ_{LV} and γ_{SL}) acting on the three phase contact line. The contact angle is determined by the energetics at the contact line, not the contact area. So it comes as no surprise that there were once serious debates about the "right-or-wrong" of the Wenzel and Cassie–Baxter analyses [3–8]. Gao and McCarthy [3] reported a very simple experimental design to test the validity of Eq. (4.4). The test samples were prepared by first creating a textured spot of diameter d on a clean silicon wafer, followed by a fluorosilane treatment. The finished samples end up with a superhydrophobic spot within the hydrophobic silicon surface. Three spot diameters, 1, 1.5, and 2 mm, were fabricated. A schematic of the heterogeneous sample structure is given in Fig. 4.1.

Experimentally, a small droplet of water was dispensed to the center of the textured area. The diameter of the contact area D was recorded and the advancing/receding contact angles θ_A/θ_R were determined. Small amount of water was added next. The diameter of the expanded drop and θ_A/θ_R of the drop were determined. This procedure is repeated for other samples, and the data are summarized in Table 4.1.

Also included in Table 4.1 are the area fractions f_1 and f_2 calculated based on the dimensions of the spot and the contact area as well as the calculated θ_A/θ_R values from Eq. (4.3). It is important to note that the calculated θ_A/θ_R value varies as the solid-area fractions for the two surface components vary. For instance, the calculated θ_A/θ_R would decrease as f_1 (area fraction of the textured area) decreases. When $d > D$ such as the cases of Exp't # 1, 5, and 9, the θ_A/θ_R values indicate that the textured area is superhydrophobic. On the other hand, when $d < D$, there is disagreement between the observed and the calculated values. Since the calculated value is entirely based on surface free energy consideration, it decreases gradually as f_1 decreases. Experimentally, all cases with $d < D$ give identical θ_A/θ_R values, indicating that the contact areas for Exp't # 2–4, 6–8, and 10–12 are all hydrophobic. The identical θ_A/θ_R values coupled with the lack of correlation with the surface energetics of the contact area led Gao and McCarthy to conclude that it is the energetics at the contact line, not the contact area beneath the drop determines the contact angle [3]. While one may argue that the Cassie–Baxter equation is still useful in

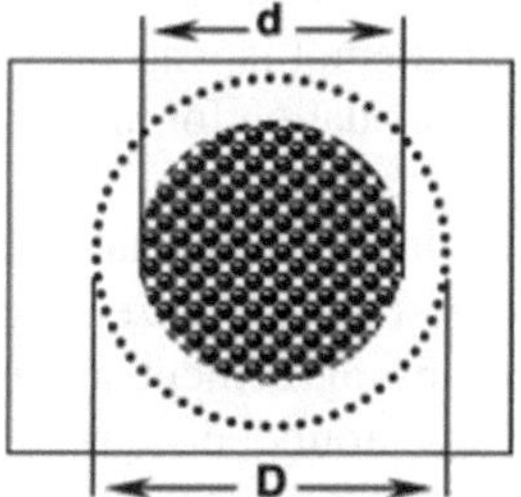

Fig. 4.1 Schematic of the heterogeneous surfaces fabricated on Si-wafer to test the Cassie–Baxter equation (*d* diameter of the texture, *D* diameter of the contact area during water contact angle measurement) (Reproduced with permission from [3], Copyright 2007 American Chemical Society)

Table 4.1 Physical and water contact angle data for the heterogeneous surfaces shown in Fig. 4.1 (data from [3])

Exp't #	d (mm)	D (mm)	f_1[a]	f_2[b]	θ_A/θ_R (cal'd)[c]	θ_A/θ_R
1	1	0.5	1.00	0.00		168°/132°
2	1	1.1	0.83	0.17	152°/122°	117°/81°
3	1	1.2	0.69	0.31	145°/115°	117°/82°
4	1	1.3	0.59	0.41	140°/108°	117°/81°
5	1.5	0.7	1.00	0.00		166°/134°
6	1.5	1.6	0.88	0.12	156°/125°	117°/82°
7	1.5	1.7	0.78	0.22	150°/119°	117°/81°
8	1.5	1.8	0.69	0.31	145°/115°	117°/82°
9	2	0.7	1.00	0.00		165°/133°
10	2	2.1	0.91	0.09	158°/126°	117°/82°
11	2	2.2	0.83	0.17	153°/122°	117°/81°
12	2	2.3	0.76	0.24	148°/118°	118°/82°

[a]f_1 area fraction of the textured area within the contact area
[b]f_2 area fraction of the non-textured area within the contact area
[c]Calculated from Eq. (4.3)

predicting the contact angle when the energetics between the contact line and the contact angle is the same. This is actually a weak argument with many flaws. This argument violates several related basic concepts that are the foundations of surface science. These basic concepts are: (1) the angle of contact between a solid and the wetting liquid is a result of a mechanical equilibrium for the three forces acting at the three phase contact line, not thermodynamic equilibrium, (2) contact angle is a one-dimensional, not two-dimensional phenomenon, and (3) surface tension and surface energy cannot be used interchangeably, the former is a tensor and the latter is scalar quantity without directional property. These basic concepts are not new. They have been discussed in fragments on-and-off from 1945 to 2010, e.g., by Pease [9], Bartell and Shepard [10], Johnson [11], Gray [12], Extrand [13], Gao and McCarthy [3, 8], and Bormashenko [14]. Although Gao and McCarthy did make an attempt to put their thoughts together and share it in their 2009 "Wetting 101°" paper [15], that effort appeared in vain as little has changed in the scientific community. We feel that this subject matter is so crucial to the future development of surface science that a second attempt is warranted. It is our hope that this work will play a role in laying a solid foundation for the basic concepts in surface science for years to come.

Together with additional results discussed later in this chapter, we agree with Gao and McCarthy that the uses of the Wenzel and Cassie–Baxter equations to predict contact angles and getting agreement are just fortuitous. In fact, many research groups [3, 16–19] have found disagreements between the calculated Wenzel and Cassie–Baxter angles with the experimentally measured contact angles, some of which will be further discussed below. We have to emphasize that we are by no means undermining the work of Wenzel and Cassie–Baxter. After all, these investigators did advance the knowledge of wetting rough surfaces and porous surfaces in the 30s and 40s, and their work had great influence to the field

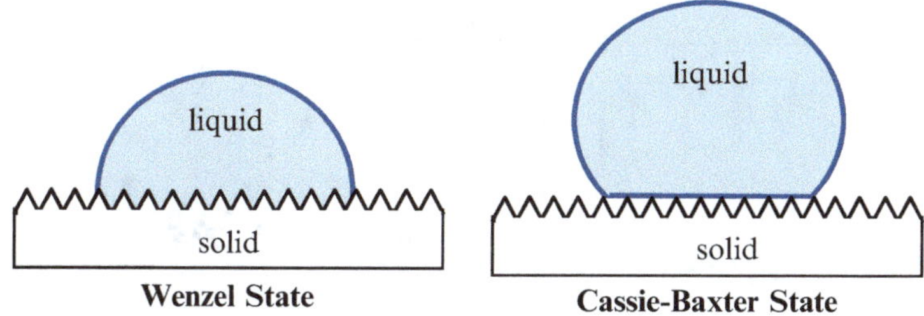

Fig. 4.2 Schematic of the two possible wetting states on rough surfaces

of surface science many decades after that. We should consider that recognizing and accepting the shortfall of the Wenzel and Cassie–Baxter analyses by itself is an important progress in surface science. Fundamentally, when a liquid wets a rough surface, there are two possible wetting states, one with the liquid fully wetting every area of the rough surface and the other with the liquid partially wetting and pinning on the asperities of the rough surface. The latter results in the formation of air pockets and a liquid–solid–air composite interface. These two wetting states have appropriately been recognized as the Wenzel and Cassie–Baxter state, respectively, in the literature (Fig. 4.2).

4.2 Wetting in the Wenzel State

According to the Wenzel equation Eq. (4.1), rough surface made of hydrophilic material can render itself superhydrophilic with a very small water contact angle θ of <10°. Rough surfaces with moderately hydrophilic and hydrophobic materials, on the other hand, can result in surfaces with large contact angles. Conceptually, when a surface is fully wetted by a liquid, it implies that the wetting process is energetically favorable. Crucial questions remain and they are: will the liquid droplet be in the most stable wetting state? What determines the contact angle? How does the contact line look like? What are the factors that control the movement of liquid on the rough surfaces? These questions will be addressed in the following.

4.2.1 The Metastable and Most Stable Wetting State

In 2004, Meiron, Marmur, and Saguy [16] reported the fabrication of four rough surfaces by coating beeswax on a glass slide and three abrasive papers (attached to glass slides) of roughness ranging from Wenzel roughness factor r 1.03–1.25 as determined by the grit number of the abrasive paper. The surface energy of these surfaces should be very similar as they were prepared from the same beeswax. The contact angles of water and ethylene glycol on these four surfaces were studied before and after vibration with a loudspeaker. The product fA represents the velocity

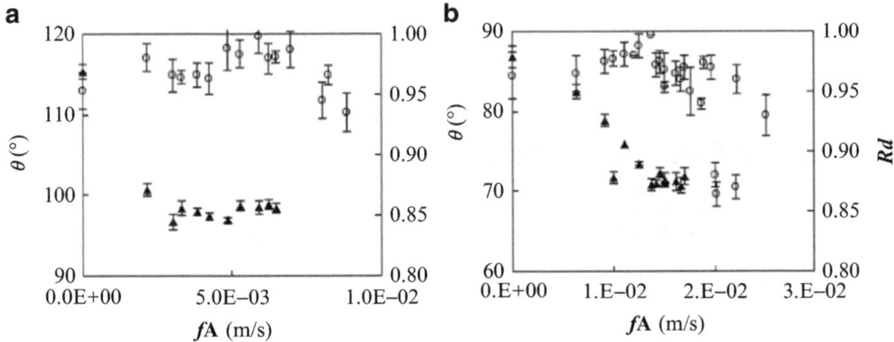

Fig. 4.3 Plot of contact angle (θ) and drop roundness (**Rd**) as a function of vibration velocity fA for surface with $r = 1.09$ (**a**) water and (**b**) ethylene glycol (*open circle* before and *filled triangle* after vibration) (Reproduced with permission from [16], Copyright 2004 Elsevier)

of the vibration motion generated by the loudspeaker. Plots of the effect of fA on the contact angle and roundness (**Rd**) of the droplets for water and ethylene glycol are shown in Fig. 4.3 for surface with $r = 1.09$. Typically, drops with **Rd** > 0.95 are considered round.

The results clearly show that vibration of the droplets using the loudspeaker de-pins the contact lines and brings the drops to the more stable wetting states with smaller contact angles. The Wenzel contact angles (θ_W) for water and ethylene glycol initially at ~115° and ~85°, respectively, were shown to decrease and subsequently level off at ~98° and ~71°, respectively, after vibration. The results suggest that the initial sessile droplets for both water and ethylene glycol are metastable. Appropriate vibration brings the metastable wetting states to their respective most stable wetting states with equilibrium angles at θ_W^{eq}. A schematic showing the free energy relationship between the initial contact angle θ_W and the equilibrium contact angle θ_W^{eq} is given in Fig. 4.4.

This free energy relationship is actually supported by the data in Fig. 4.3. For instance, the viscosity for ethylene glycol is higher than that of water, suggesting that the energy barrier for the liquid advance is higher for ethylene glycol on the same rough surface. Indeed, comparison of the fA values in Fig. 4.3a, b reveals that it takes about ten times more energy to bring the ethylene glycol droplet to the equilibrium state as compared to that of water. Table 4.2 compares the observed θ_W^{eq} values for water and ethylene glycol on the four rough surfaces with the calculated Wenzel angle $(\theta_W)^{cal}$ values from the Wenzel equation Eq. (4.1). A very good agreement is obtained. The agreement confirms that, as with smooth surface, the apparent contact angle θ_W on rough surface is from a metastable wetting state. Liquid droplet just ceases to spread as all of its kinetic energy is dissipated due to friction created by the rough surface. This results in a larger than expected θ_W value. When the metastable wetting state is excited by the vibration noise, the contact line de-pins and continues to advance to the most stable wetting state with an equilibrium angle θ_W^{eq}.

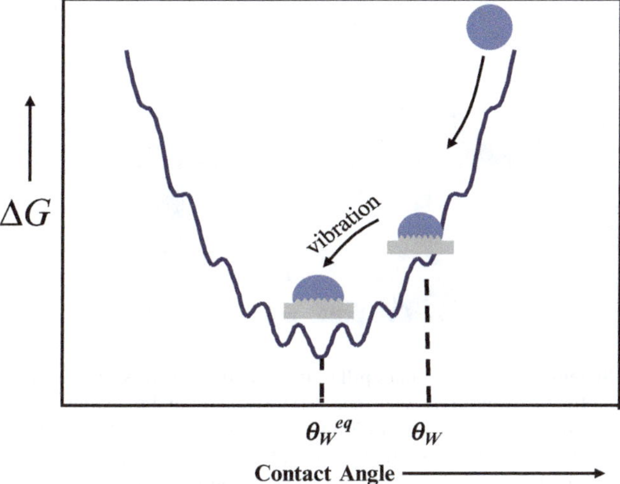

Fig. 4.4 Schematic of the free energy relationship between Wenzel angle θ_W and the equilibrium contact angle θ_W^{eq} on rough surface

Table 4.2 Comparison of equilibrium Wenzel angle θ_W^{eq} with the calculated Wenzel angle $(\theta_W)^{cal}$ (data taken from [16])

	Water		Ethylene glycol	
r	θ_W^{eq}	$(\theta_W)^{cal}$	θ_W^{eq}	$(\theta_W)^{cal}$
1.01	97.3	97.3	73.1	73.2
1.03	97.9	97.7	72.4	72.9
1.09	98.0	97.3	71.6	73.2
1.25	100.1	98.1	69.7	73.9

4.2.2 Unexpected Wettability

The Wenzel equation suggests that roughness should increase the wettability of hydrophilic surface. This is certainly true for highly hydrophilic materials (water $\theta < 30°$), where roughness has often led to super-wetting or superhydrophilicity [20]. On the other hand, the wetting behavior of rough surface made of moderately hydrophilic material has not been well studied. In 2010, Forsberg and co-workers [17] reported a study of the wetting of microtextured surfaces comprising square pillar arrays made of SU8. SU8 is a common photoresist material used in photolithography. It is moderately hydrophilic with a water θ of ~72°. Figure 4.5a, b show the schematics of the pillar array design. The width of the square pillar is fixed at $w = 20$ μm, the pitch d varies from 25 to 120 μm, and the heights studied are at 7 and 30 μm. Figure 4.5c shows the water advancing and receding angle of the control smooth SU8 surface, which is at 72° and 59°, respectively. On the pillar array surface shown in Fig. 4.5d ($w/d = 0.63$, height 7 μm), an unexpected large advancing contact angle at 140° was observed. The surface exhibits practically 0° receding angle and a photograph showing the meniscus after water receding is shown in the insert. The result suggests that water fully wets the SU8 pillar array surface. Additional control experiment was performed with a hydrophobized pillar array

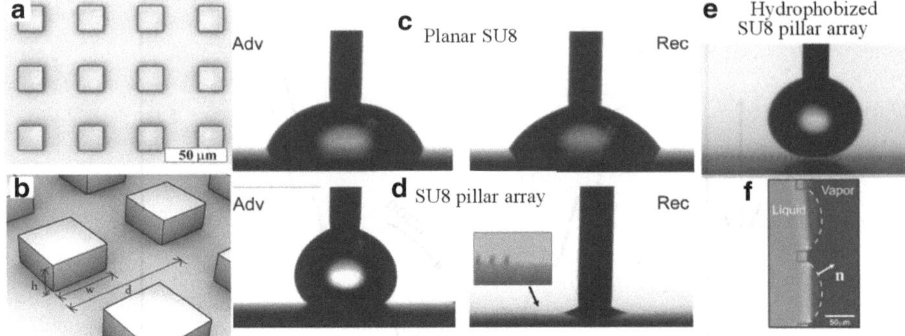

Fig. 4.5 (**a**, **b**) Schematics for the square pillar array model surfaces, (**c**) advancing and receding contact angle data for the smooth SU8 surface, (**d**) advancing and receding contact angle data for an SU8 pillar array surface (w/d=0.63, h=7 µm), (**e**) water sessile drop data on a hydrophobized SU8 pillar array surface, and (**f**) photograph of the advance contact line (d=120 µm) (Reproduced with permission from [17], Copyright 2010 American Chemical Society)

Fig. 4.6 Plot of cosθ versus w/d for SU8 square pillar array surfaces (Reproduced with permission from [17], Copyright 2010 American Chemical Society)

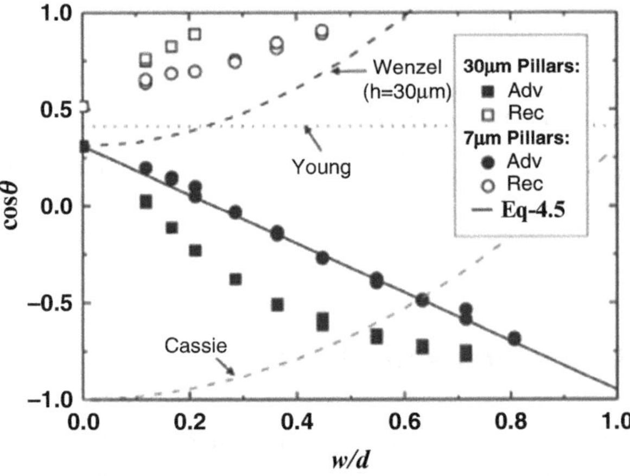

surface. The sessile drop data is given in Fig. 4.5e. A large static contact angle with small hysteresis was obtained, indicating that the water droplet is indeed in the Cassie–Baxter state on the hydrophobized pillar array surface. The overall wetting data allow one to conclude that water droplet is in the Wenzel state on the SU8 pillar array surfaces.

Figure 4.6 plots the advancing and receding contact angles of the SU8 pillar array surfaces as a function of w/d. The upper dash line shows the trend calculated from the Wenzel equation Eq. (4.1). The observed advancing contact angles and the calculated data are not only in disagreement, they are actually trending in an opposite direction. A closer examination of the advancing contact line on the pillar array surface (Fig. 4.5f) shows that the contact line pins at the corners of the square and extends outward in the space between pillars. To explain the unexpected large contact angle and the disagreement between the Wenzel equation and the experimental data, Forsbery and co-workers [17] first assumed that the liquid–solid–air interface is in a mechanically stable configuration. They then used the dimension and geometry of the contact line to modify the Wenzel roughness factor. The modified Wenzel equation is given by:

$$\cos \theta_{\mathrm{W}}^{\mathrm{mod}} = \left(1 - \frac{w}{d}\right) \cdot \cos \theta_{\mathrm{SU8}} + \left(\frac{w}{d}\right) \cdot \cos\left(\theta_{\mathrm{SU8}} + \frac{\pi}{2}\right) \tag{4.5}$$

The $\cos \theta_{\mathrm{W}}^{\mathrm{mod}}$ values were calculated and shown as the solid line in Fig. 4.6. A very good agreement is obtained between the modified Wenzel angle and the observed angle for the pillar array surfaces with a 7 μm pillar height. For pillar array surfaces of 30 μm pillar height, the observed apparent contact angles were even larger. The discrepancy with the calculated values from Eq. (4.5) is attributable to the assumption of using a low pillar height in the modified equation. In any event, this study again supports the notion that contact angle is governed by where the contact line is pinned, not the contact area beneath the liquid droplet.

The experimental observation by Forsbery and co-workers is consistent with the wetting dynamics on solid surfaces as discussed in Chap. 3. For instance, as water starts wetting the 7 μm pillar array surface, the contact line advances. The advance will cease when all its kinetic energy is dissipated. Due to the large friction created by the pillar array, the advance contact line is stopped far from its equilibrium position, resulting in an unexpected large contact angle, consistent with the free energy model shown in Fig. 4.4. This interpretation gains further support from the data collecting from the 30 μm pillar height, pillar array surfaces. Due to the increase in pillar height, the energy barrier or friction against wetting is larger for the 30 μm pillar array surfaces. The analysis would suggest that the advancing contact angles for the 30 μm pillar height surfaces would be larger than those of the 7 μm pillar height surfaces. Indeed, this is experimentally observed.

4.2.3   Roughness Geometry on Wettability and Wetting Dynamics

A series of model rough PDMS surfaces comprising arrays of 3 μm hemispherical bumps and cavities with pitches ranging from 4.5 to 96 μm were recently fabricated by Kanungo and co-workers using the conventional photolithography and molding techniques [19]. The representative SEM micrographs are given in Fig. 4.7. These surfaces are designed to address the following questions: are surfaces with bumps and cavities better models to emulate wetting and de-wetting behavior of liquid on real, practical rough surfaces, which are mostly hills and valleys? The wetting entrant angles between real rough surfaces and the bumpy and cavity surfaces shown in Fig. 4.7 are fairly comparable, significantly less than 90°. Whereas majority of the model rough surfaces reported in the literature are pillar arrays with high

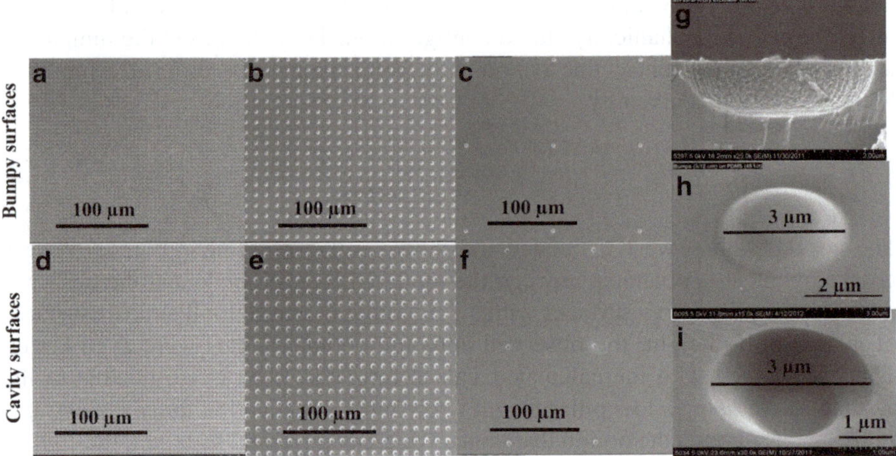

Fig. 4.7 Representative SEM micrographs of rough PDMS surfaces with varying pitches (**a, d** 4.5 μm, **b, e** 12 μm and **c, f** 96 μm); (**g**) SEM micrograph of the 3 μm diameter hemispherical silicon mold, (**h**) SEM micrograph of the hemispherical bump on the PDMS surface created from the mold in (**g**), and (**i**) SEM micrograph of the hemispherical cavity PDMS surface molded from the bumpy surface in (**g**) (Reproduced with permission from [19], Copyright 2014 American Chemical Society)

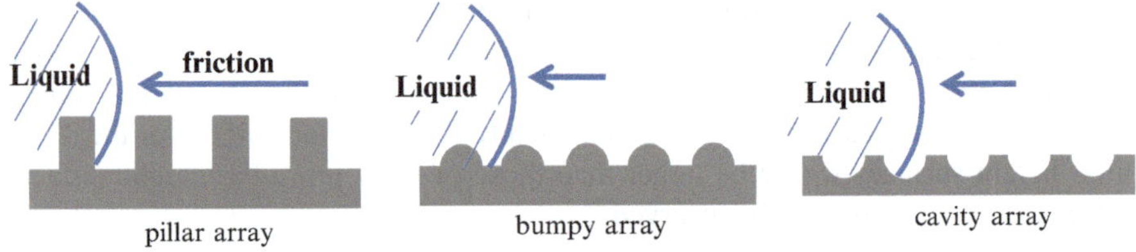

Fig. 4.8 Schematic of the wetting scenarios between a pillar array rough surface and the hemispherical bumpy and cavity array surfaces

aspect ratios [17, 21–23]. Pillars are vertical protrusions from a flat surface and the wetting entrant angle against the advancing liquid is at 90°. The friction the pillars exerted to the advancing liquid is expected to be larger than those from hills and valleys and also from bumps and cavity surfaces based on simple geometrical consideration. Schematics showing the differences in frictions created by these surfaces are shown in Fig. 4.8. Other objectives of the work include study of the effect of rough geometry, using bumps and cavities as models for hills and valleys, on surface wettability and wetting dynamics.

Wetting States on Model Rough PDMS Surfaces. The surface properties of all model bumpy and cavity PDMS surfaces were studied by static and dynamic contact angle measurements with water as the test liquid and the data are tabulated in Table 4.3. The contact angle data for the smooth PDMS surface are included as reference. As discussed in Sect. 4.1, wetting of liquid on a rough surface can be described by

Table 4.3 Contact angle measurement data for model rough PDMS surfaces with arrays of bumps and cavities (Reproduced with permission from [19], Copyright 2014 American Chemical Society)

Pitch[a]	r[b]	θ_W[c]	θ_{CB}[d]	Array of bumps				Array of cavities				r^{eff}[i]	θ_W^{mod}[j]
				θ[e]	θ_A[f]	θ_R[g]	CAH[h]	θ[e]	θ_A[f]	θ_R[g]	CAH[h]		
96 µm	1.00	110°	110°	116°	117°	69°	48°	117°	120°	76°	44°	1.05	111°
48 µm	1.01	110°	110°	115°	119°	71°	48°	117°	120°	77°	43°	1.11	112°
24 µm	1.02	111°	110°	117°	120°	68°	52°	118°	119°	79°	40°	1.22	115°
12 µm	1.1	112°	112°	122°	125°	61°	64°	124°	129°	81°	48°	1.44	119°
6 µm	1.4	118°	118°	129°	131°	68°	63°	128°	132°	83°	49°	1.87	130°
4.5 µm	1.7	126°	125°	138	144°	83°	61°	136°	140°	88°	52°	2.16	138°
Smooth surface	1.00	–	–	110°	112°	72°	40°	–	–	–		–	–

[a]Center-to-center spacing between bumps or cavities
[b]The classic Wenzel roughness ratio, actual surface area divided by projected area
[c]Calculated Wenzel angle from Eq. (4.1)
[d]Calculated Cassie–Baxter angle (θ_{CB}) for cavity surface from Eq. (4.2)
[e]Static contact angle
[f]Advancing contact angle
[g]Receding contact angle
[h]Contact angle hysteresis, defined as ($\theta_A - \theta_R$)
[i]Effective roughness factor calculated by correcting for the increase of contact line density at the three-phase contact line [17]
[j]Modified Wenzel angle calculated according to method of Forsberg and co-workers [17]

the two classic wetting models. When the rough surface is fully wetted by the liquid, the apparent static contact angle is the Wenzel angle (θ_W) and is given by the Wenzel equation [1]. The Wenzel roughness ratios (r) for the model rough PDMS surfaces can be calculated from the radii and geometry of the bumps and cavities. The r values are the same for the bumpy and cavity surfaces of the same pitch. Details of the calculation have been given elsewhere [19]. From the r values, the Wenzel angles (θ_W) are calculated and are given in Table 4.3 column 3.

For rough, porous surface where pockets of air can be created during wetting, a composite liquid–solid–air interface is formed. The apparent contact angle is the Cassie–Baxter angle (θ_{CB}) and is given by the Cassie–Baxter equation [2]. Figure 4.9 shows the generalized, hypothetical wetting states for (a) the bumpy PDMS surface and (b) the cavity PDMS surface. On the bumpy PDMS surface, h_L is the height needed to create the air pocket. On the cavity surface, h_L is the sagging height. Since optical microscopy results (Fig. 4.10 below) indicate that the contact lines for both bumpy and cavity surfaces are pinned at the lead edge of the rough structures, that means $h_L = 0$ for both the bumpy and cavity surfaces. It is thus geometrical impossible to trap air on the bumpy surface when $h_L = 0$ during wetting. In the other word, bumpy PDMS surfaces will always be in the Wenzel wetting state. However, a Cassie–Baxter state is still a possibility for the cavity surface when $h_L = 0$. The Cassie–Baxter angles (θ_{CB}) for cavity surfaces are then calculated (Table 4.3 column 4).

The overall results in Table 4.3 show that the observed θ for the model rough PDMS surfaces (columns 5 and 9) are not in agreement with neither θ_W nor θ_{CB}

Fig. 4.9 Generalized, hypothetical schematics for the wetting of (**a**) the bumpy PDMS surface and (**b**) the cavity PDMS surface. On the bumpy PDMS surface, h_L is the height needed to create the air pocket. On the cavity surface, h_L is the sagging height (Reproduced with permission from [19], Copyright 2014 American Chemical Society)

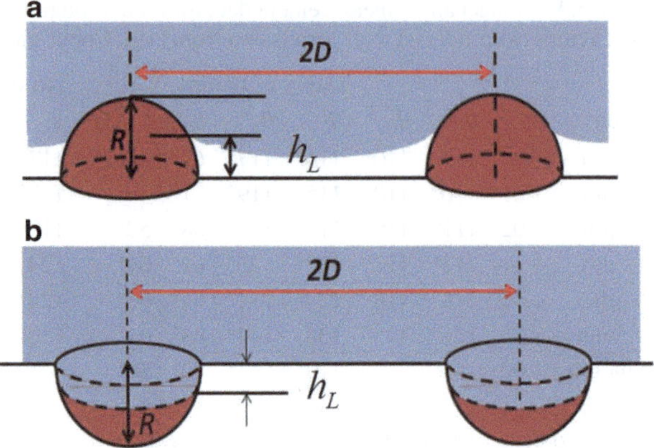

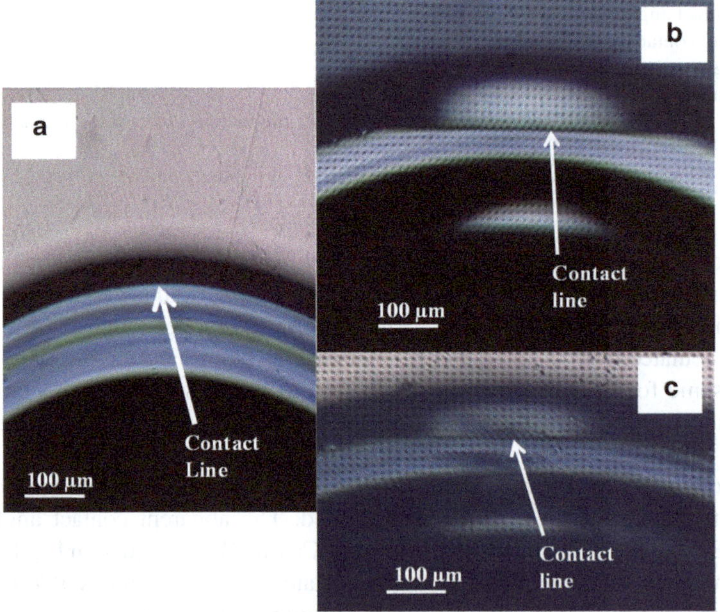

Fig. 4.10 Optical photographs of the three-phase contact lines as imaged from the bottom of the water sessile droplets on (**a**) smooth, (**b**) 12 µm pitch bumpy, and (**c**) 12 µm pitch cavity PDMS surfaces (Reproduced with permission from [19], Copyright 2014 American Chemical Society)

values. The disagreement is not surprising as recent theoretical and experimental results showed that the contact angles of rough surfaces tend to correlate more to the locality of the three phase contact line than the classic Wenzel and Cassie–Baxter angles due to pinning of the contact lines on rough surfaces [17, 18, 24–27]. Using the methodology of Forsberg and co-workers [17], a one-dimensional modified

Wenzel equation can be derived based on the geometry of the contact line (Fig. 4.10), where θ_W^{mod} is the modified Wenzel angle.

$$\cos \theta_W^{mod} = \left(1 - \frac{R}{D}\right) \cdot \cos \theta + \left(\frac{R}{D}\right) \cdot \cos\left(\theta + \frac{\pi}{2}\right) \tag{4.6}$$

After rearranging the above equation to the classical Wenzel format, the effective roughness ratio (r^{eff}) is obtained and is given by:

$$r^{eff} = 1 - \frac{R}{D}\left(1 + \tan \theta\right) \tag{4.7}$$

Details for the derivations of r^{eff} and θ_W^{mod} and discussion of the wetting states have been given earlier [19]. The calculated r^{eff} and θ_W^{mod} values are listed in Table 4.3 columns 13 and 14, respectively. Without exception, reasonably good agreements between θ_W^{mod} and θ are observed for both bumpy and cavity surfaces, suggesting that both type of surfaces are in the fully wetted Wenzel states. This conclusion is supported by recent wetting studies of PDMS pillar array surfaces both theoretically and experimentally. Jopp et al. [28] showed by free energy calculation that water will fill all the grooves between pillar arrays of PDMS, which is hydrophobic. Papadopoulos and co-workers [22] reported visualization of the fully wetted liquid–solid interface between water and PDMS three dimensionally by laser scanning confocal microscopy. Additional evidence supporting the conclusion that water is in the fully wetted Wenzel state on both bumpy and cavity PDMS surfaces come from the drop vibration experiments, which will be given later in this chapter.

Advancing and Receding Contact Lines The location and the geometry of the advancing contact lines on the bumpy and cavity PDMS surfaces were examined directly from the bottom of the water droplets and the optical micrographs of the contact lines are given in Fig. 4.10. Expectedly, the contact line for the smooth PDMS surface is smooth and round (Fig. 4.10a), whereas those on the bumpy and cavity surfaces are distorted by the microstructures (Fig. 4.10b, c). This is consistent with the contact lines observed in other microtextured surfaces, where the three-phase contact lines are all shown to follow the edge of the rough microstructures [18, 21, 22, 24–27]. A closer examination of the photographs reveals that the three-phase contact lines in Fig. 4.10 actually follow the lead edges of the bumps on the bumpy surface and the cavities on the cavity surface. A schematic showing the top-view and side-view of the three-phase contact lines on these surfaces is given in Fig. 4.11. Both contact lines are shown pinning at the lead edges of the rough structures.

Surfactants and dyes in aqueous solutions are known to localize around the three-phase contact lines [29]. As water is evaporated, the dye crystalizes and the residues provide a trace for the receding contact line. Using the procedure analogous to that reported by Wu and co-workers [30], the receding contact lines were imaged by evaporating the sessile droplets of a very dilute Rhodamine solution ($\sim 5 \times 10^{-6}$ g/mL) on the smooth and the bumpy and cavity PDMS surfaces. Controlled experiments show that the added dye in water has no effect on the contact angles on all PDMS surfaces. Figure 4.12a–c show the optical photographs of the evapo-

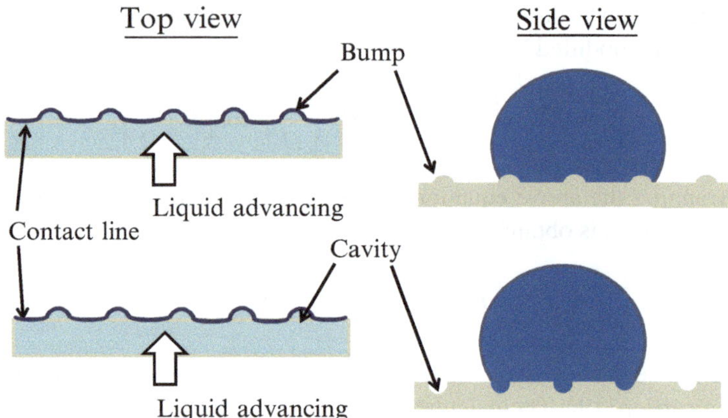

Fig. 4.11 Schematics for the top-view and side-view of the three-phase contact lines for water droplets on the bumpy and cavity PDMS surfaces (Reproduced with permission from [19], Copyright 2014 American Chemical Society)

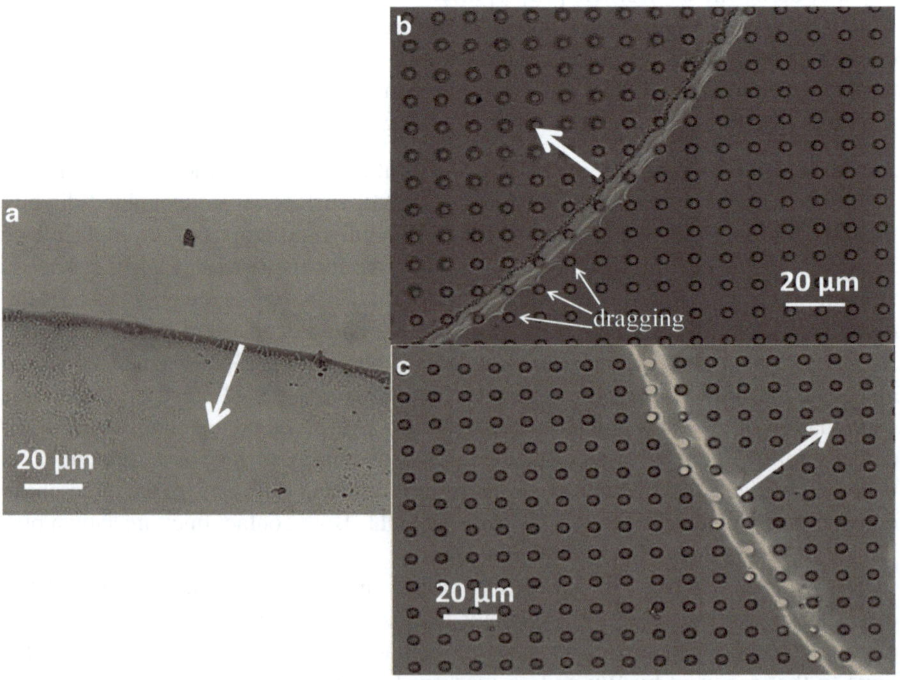

Fig. 4.12 Optical photographs of the receding contact lines as imaged by evaporating a very dilute Rhodamine dye solution (~5 × 10^{-6} g/mL) on (**a**) smooth, (**b**) 12 µm pitch bumpy, and (**c**) 12 µm pitch cavity PDMS surfaces. The arrows show the receding directions (Reproduced with permission from [19], Copyright 2014 American Chemical Society)

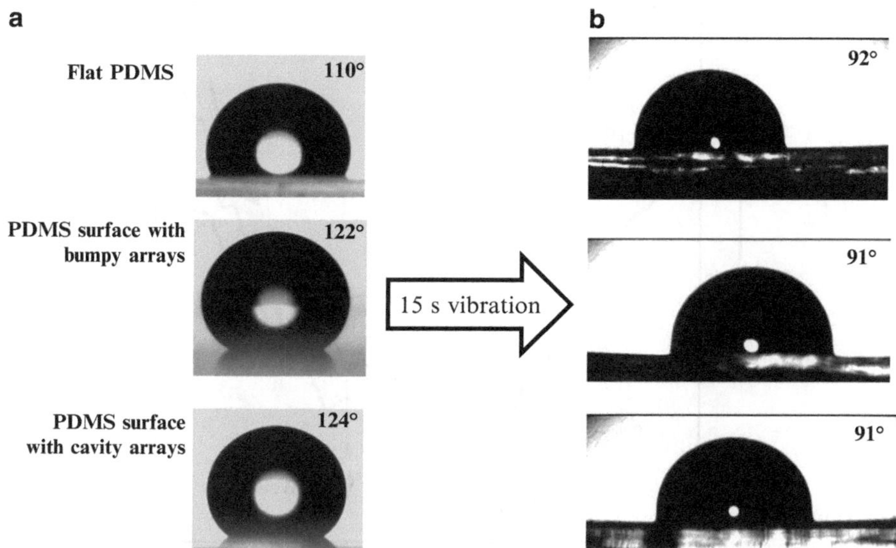

Fig. 4.13 Images of the water sessile droplets on smooth and rough PDMS surfaces (**a**) before and (**b**) after 15 s vibration (Reproduced with permission from [19], Copyright 2014 American Chemical Society)

rated droplets on the smooth PDMS surface and the bumpy and cavity PDMS surfaces, respectively. The photograph in Fig. 4.12a clearly shows the dye residue on the smooth PDMS surface after water evaporation. The contact line is round and smooth during the liquid-receding process. On the other hand, microstructures are visible in the receding contact line on the bumpy PDMS surface (Fig. 4.12b). The image suggests that the bumps are dragging the contact line as the liquid is receding. As for the cavity surface, the receding contact line is relatively smooth and round. There appears to be little dragging by the cavities. The microscopy results thus indicate that rough geometry does have an effect on the movement of the contact line. Bumps are found to drag the contact line as it recedes while cavities behave like a smooth surface.

Drop Vibration Experiments Fig. 4.13a depicts the photographs of the ~5 μL sessile water droplets on the smooth PDMS surface, the 12 μm pitch bumpy PDMS surface, and the 12 μm pitch cavity PDMS surface. Their static contact angles are at 110°, 122°, and 124°, respectively. To test the wetting states of the water on these surfaces, Kanungo and co-workers [19] subjected all three water droplets in Fig. 4.13a a 15-s vibration according to the procedure described by Cwikel and co-workers [31] and the results are given in Fig. 4.13b. All three sessile droplets give the same static contact angles at ~91° after the 15-s vibration. The results indicate that all three sessile droplets were in the metastable wetting states before the vibration. While the static contact angle for the smooth PDMS surface changes from 110° to 92° after vibration and the observation is consistent with that reported

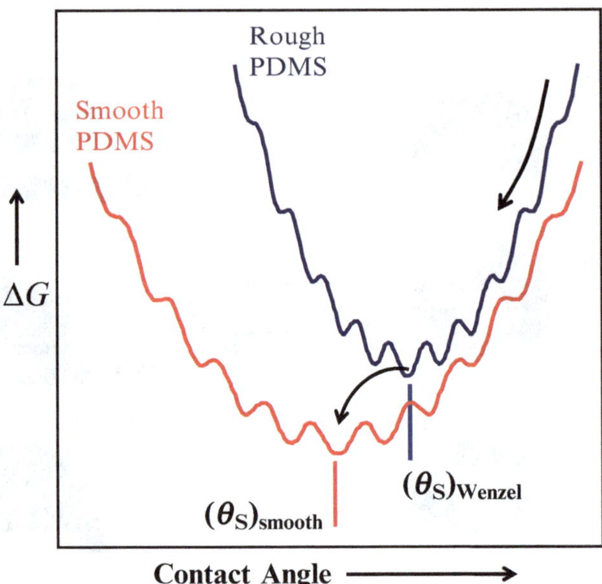

Fig. 4.14 Schematic plot of the Gibbs free energy curves for wetting of a smooth PDMS surface and rough PDMS surface (bumps and cavities) by water as a function of the apparent contact angle. θ_S is the most stable contact angle (Reproduced with permission from [19], Copyright 2014 American Chemical Society)

by Cwikel and co-workers [31], the results for the bumpy and cavity surfaces are very unusual as their contact angles after vibration would suggest that the water droplets pin selectively in the smooth areas of the rough PDMS surface! Indeed, further optical microscopy study reveals that, the contact lines on both bumpy and cavity PDMS surfaces become much smoother after vibration and that they appear to be on the smooth area of the PDMS surfaces rather than distorted by the rough structures [19]. The overall results from the drop vibration experiments suggest that water droplets are always in the metastable wetting states on the bumpy and cavity surfaces due to pinning effect. Vibration, which leads to de-pinning and relocation of the contact lines, results in the population of the most stable wetting states on these surfaces. The similar final θ values for all the PDMS surfaces after vibration suggest that the most stable wetting states on the bumpy and cavity surfaces also involve pinning the water droplets in the smooth area of the rough PDMS surfaces. The observation can be rationalized based on the Gibbs free energy curves of the wetting states of the smooth and rough surfaces. The schematics of the curves are given in Fig. 4.14 based on the contact angle data in Table 4.3. As discussed in Chap. 3 and Sect. 4.2, wetting a solid surface involves transfer of the kinetic energy in the sessile droplet (gained from gravity acceleration) to the kinetic energy for wetting after the liquid droplet contacts the surface. The droplet will cease to spread when all its kinetic energy is dissipated. The generally large θ values observed in Table 4.3 for the bumpy and cavity PDMS surfaces relative to the calculated Wenzel angles (θ_W) can thus be attributed to this pinning effect [17, 22, 26]. In the case of a fully wetted water droplet on a rough surface, Meiron and co-workers [16] showed that vibration can convert a metastable wetting state to the most stable Wenzel state. The contact angle for the stable Wenzel state was found to be in agreement with the calculated Wenzel angle (θ_W). In the cases of the model PDMS surfaces in the Kanungo study, contrast to that reported by Meiron and co-workers, the most stable contact angles for the bumpy and cavity surfaces are a lot smaller than the calculated θ_W (110°–126° vs. ~91°). The similar contact angles for all three PDMS surfaces after the 15-s vibration suggest that the contact lines on the rough surfaces relocate to the smooth area of the PDMS surface after de-pinning by vibration. This is supported by recent optical microscopy results and implies that the stable Wenzel states for the bumpy and cavity surfaces are not as thermodynamically favorable as that of the smooth surface. We suggest that during vibrational excitation, the contact lines de-pin and the droplets on the bumpy and cavity surfaces are able to cross-over to the energy curve of the smooth PDMS surface (Fig. 4.14).

Effects of Pitch Length and Geometry on Wettability. Results in Table 4.3 show that the static contact angles (θ) increases as pitch length decreases for both types of rough surfaces (bumps and cavities) due to the increase in roughness on the PDMS surface. It is worthy pointing out that for the same pitch, where the roughness ratio is the same, the static contact angles for the bumpy and cavity surface are actually very similar.

Interesting results are observed for the dynamic contact angle data. The advancing contact angle (θ_A) for the bumpy surface and the cavity surface are comparable for the same pitch and they increase, from ~116° to ~138°, as r^{eff} increases from 1.05 to 2.16. Although similar trend is also observed for the receding contact angles (θ_R) (increase as r^{eff} increases), the receding angles for the bumpy surfaces are consistently smaller than those of the cavity surfaces. In the other word, for the same pitch, the hysteresis for the bumpy PDMS surface (column 8) is always larger than that of the cavity surface (column 12). Similar asymmetric hysteresis has been observed for pillar and porous array surfaces by Priest et al. [21]. In the case of the bumpy and cavity PDMS surfaces, Kanungo et al. actually have visual evidence to show that bumps are exerting strong resistance and drag the receding contact line (Fig. 4.12b) whereas minimal dragging is seen by the cavities (Fig. 4.12c). These authors thus attribute the asymmetric hysteresis observed to the dragging of the contact line by the bumps as the liquid recedes [19].

4.2.4   Practical Consequences

Being able to wet every cavity or hole in a rough surface has advantages in both nature and man-made devices. When water wets a superhydrophilic surface, it spreads spontaneously with no trap air and no measurable contact angle. In nature, such surface structures have been created in plant leaves so that the plants can adapt

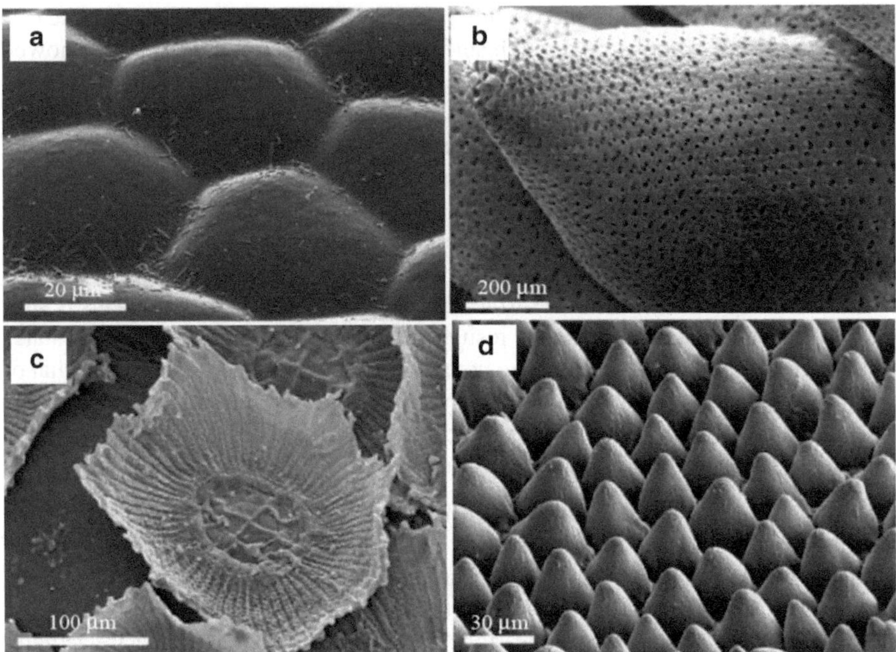

Fig. 4.15 SEM micrographs of some superhydrophilic plant leaves (**a**) Anubias barteri, (**b**) Sphagnum squarrosum, (**c**) Spanish moss *Tillandsia usneoides*, and (**d**) *Calathea zebrina* (Reproduced with permissions from [32], Copyright 2008 The Royal Society of Chemistry and [33] Copyright 2009 Elsevier)

and survive in their perspective living environment [32, 33]. For instance, the microstructures in submerging water plant leaves tend to be smooth, enabling water to evenly wet the surface while discourage fouling (Fig. 4.15a). Contrarily, the microstructures of water-absorbing plants, such various mosses, are structurally more complex (Fig. 4.15b, c). These leaves are for uptake of water and nutrient, so the surface is designed to be superhydrophilic with a structure to hold onto the absorbed water. On the other hand, plant leaves from *Calathea zebrina* (Fig. 4.15d) will have a very simple microstructure because all it needs to do is to spread water very fast on the surface.

Owing to the super-spreading ability, researchers have envisioned that superhydrophilic surfaces should enable antifogging as minute water droplets will spread instead of pinning and fogging the surface. Similarly, the spreading action leads to wet self-cleaning as dust and dirt particles will wash away during water spreading. Many artificial superhydrophilic surfaces are known and the most notable one is the film of TiO_2 [34]. The surface is prepared by coating a sol gel precursor on a substrate followed by annealing the coated film in a furnace. TiO_2 film is intrinsically, moderately hydrophilic. Its superhydrophilicity is activated by UV or sunlight radiation. Antifogging devices and self-cleaning structures (Fig. 4.16) have been commercialized, and the subject has been reviewed in the literature [35]. Similarly to

Fig. 4.16 (a) Photograph of a foggy TiO_2 coated glass, (b) surface in (a) after sufficient UV illumination, and (c) photograph of the MM Towers in Yokohama, Japan where self-cleaning tiles were used (Reproduced with permission from [35], Copyright 2008 Elsevier)

TiO_2, silica and other metal oxide particles are hydrophilic too. They can be incorporated into thin films via the sol gel technique [36], or by layer-by-layer deposition [37] or as a composite coating with a polymer binder [38]. Finally, a fully wetted liquid–solid interface should also be beneficial for heat-exchange devices and electrodes, where overheating and over-potential can be minimized.

4.3   Wetting in the Cassie–Baxter State

4.3.1   The Lotus Effect

When a liquid is in the Cassie–Baxter state on a rough surface, the droplet is usually characterized by a large contact angle with a small sliding angle. For water, the surface is designated as superhydrophobic. The most famous superhydrophobic surface is the Lotus leaf, exhibiting a water contact angle of 162° and a sliding angle of 4° [39]. During rolling off, dust and dirt particles adhere to the water droplet, resulting in the so-called self-cleaning effect. This observation has inspired numerous researchers worldwide and studies of superhydrophobicity have grown exponentially since. Fig. 4.17 shows the SEM micrographs of a leaf surface, which comprises micron-size aggregated wax crystals randomly distributed on the ~10–20 μm

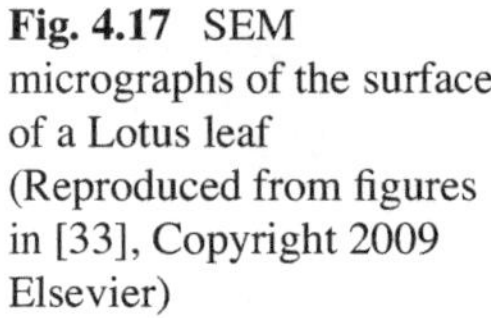

Fig. 4.17 SEM micrographs of the surface of a Lotus leaf (Reproduced from figures in [33], Copyright 2009 Elsevier)

papillae with the entire leaf surface carpeted with a layer of waxy nano-hairs (tubules) [32, 33, 39–41].

Tremendous attention has been paid to understand the hierarchical (multi-scale) surface structure and its effect on the super water repellency, hysteresis, wetting stability under pressure (e.g., against heavy raindrop), and mechanical properties [42–49]. Free energy analysis and thermodynamic modeling of hydrophobic surfaces suggest that a dual roughness scale (micron to submicron/nano) would result in large water contact angle, low contact angle hysteresis, and high wetting stability against the Cassie–Baxter to Wenzel transition. Theoretical calculations consistently reveal that a large contact angle is already achieved with a microscale surface and that the hierarchical structure is designed for low hysteresis and the ability to maintain the non-wetting state under high pressure. This is in agreement with experimental data [49]. More importantly, modellings also suggest that the micron-size feature is for mechanical stability of the surface [47]. The knowledge gained from nature has been very useful for the design and fabrication of artificial superhydrophobic surfaces.

The structure of the liquid–solid–air composite interface between water and the Lotus leaf was recently studied by Luo and co-workers using 3D confocal laser scanning microscopy, and the results are depicted in Fig. 4.18 [50]. With the bare leaf surface (Fig. 4.18a), a featureless rough dark surface is observed. Here, dark is indicative of solid surface. A very different image is observed for the interface. The image is bright indicative of reflective light by the trapped air. 3D roughness measurement of the composite interface suggests that the average thickness of the air cushion between the leaf surface and the water droplet is ~15 µm. This is inconsistent with the hierarchical structure shown in Fig. 4.17 and suggests that the main contact areas between water and the leaf surface are the tips of the papillae.

Although the chemical composition of plant wax has not been well characterized, Cheng and co-workers [40] found that the wax material on the Lotus is intrinsically, moderately hydrophilic with a water contact angle of 74°. As will be discussed in the next two sections, re-entrant structure and multi-scale roughness at

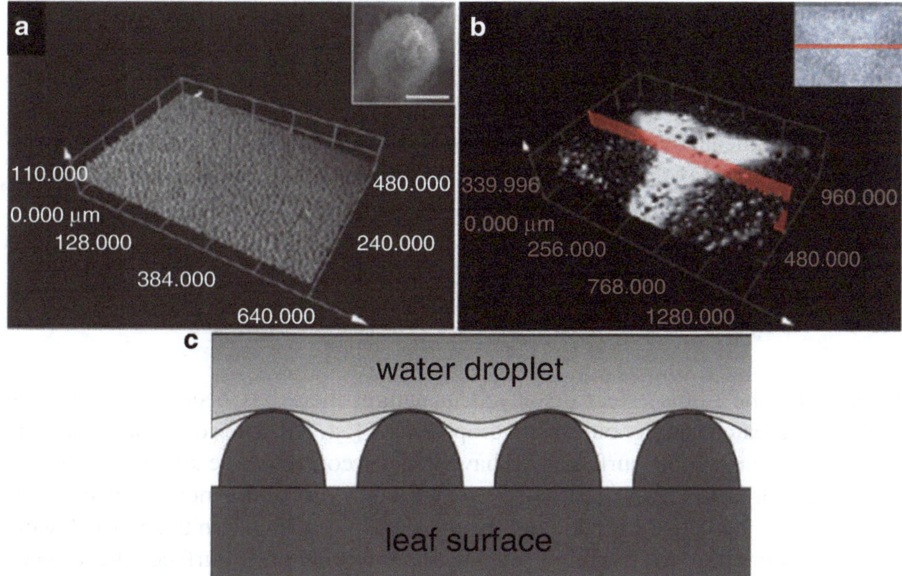

Fig. 4.18 Confocal laser scanning images of Lotus leaf (**a**) bare leaf (**b**) interface with a water drop, and (**c**) schematic of the interface from the image in (**b**) (Reproduced with permission from [50], Copyright 2010 Wiley)

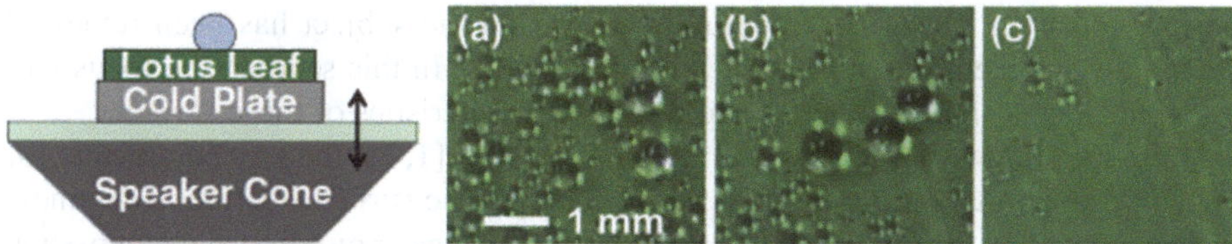

Fig. 4.19 Left: experimental setup for where the cold plate was used for the water condensation experiment; (**a**) condensation of water droplets on the Lotus leaf at cold plate temperature of 5 °C, (**b**) water droplets on horizontal Lotus leaf after titling (**a**) at 90°, and (**c**) water droplets on horizontal Lotus leaf after loudspeaker vibration of sample (**a**), followed by tilting the leaf to 90°. Most water droplets roll off after titling (Reproduced with permission from [53], Copyright 2009 American Physics Society)

the liquid–solid–air composite interface play critical roles in stabilizing the Cassie–Baxter state with moderately hydrophilic material. We believe that similar stabilization mechanism is operating when water wets the waxy tubules on the Lotus leaf [51]. The hydrophilic nature of the leaf surface is supported by water condensation experiments reported by Cheng and Rodak [52] and Boreyko and Chen [53]. Both groups showed that minute water droplets can condense on the Lotus leaf surface at high humidity. The experimental setup for the Boreyko and Chen experiment is given in Fig. 4.19. Basically, a fresh Lotus leaf is mounted onto a substrate under lab ambient condition of 21 °C at 50 % RH. The substrate is then cooled to 5 °C using a circulating bath. Minute water droplets were condensed and later coalesced into larger water droplets of ~2 mm in diameter (Fig. 4.19a). These drops are found to be in the Wenzel state as they are shown sticking on the Lotus surface when tilted to 90° (Fig. 4.19b). On the other hand, when these Wenzel drops are subjected to vibration noise from a loudspeaker (80 Hz, 1 mm peak-to-peak for 1 s), most water drops become mobile. They slide off the leaf surface upon tilting (Fig. 4.19c). This experiment indicate that, although the Cassie–Baxter state of the Lotus leaf is meta-stable, its energy level is fairly close to the most stable Wenzel state and can be populated by a simple vibration.

4.3.2   *Artificial Superhydrophobic Surfaces*

Inspired by the Lotus effect displayed in nature, there has been an exponential growth in research activity on superhydrophobicity [54, 55]. The consensus definition for a superhydrophobic surface is to have a water contact angle >150° and sliding angle ≈ 10° or less. This definition is somewhat arbitrary and a more refined definition has been published [56] and will be the subject of Chap. 6 in this book. Owing to the fascinating self-cleaning effect, superhydrophobic surfaces have been exploited for many potential applications, such as self-cleaning windows and textiles, oil- and soil-resistant clothing, anti-smudge surfaces for i-phone and display, anti-icing coatings for power line, roof-top and airplane, corrosion-resistant coatings for bridges and other metal structures, drag reduction in ship, gas and fuel transportation and microfluidic devices, etc. Numerous artificial superhydrophobic surfaces/coatings have been reported to date, and the subject has been reviewed frequently in recent literature (last 3 years) [57–62]. In this section, the focus will be on the key design parameters and wetting characteristics of these surfaces.

Based on the work of Wenzel and Cassie–Baxter [1, 2], the key parameters for superhydrophobicity are roughness and a hydrophobic surface coating. Roughness can be created by the top-down or bottom-up techniques or molding and embossing. Detailed descriptions of these processes have been reviewed in the literature and will not be detailed here. In the top-down approach, the roughness is created by etching a smooth surface with an energy (photons or electrons) beam or an etching agent, followed by modifying the surface with a hydrophobic coating if the material by itself is not hydrophobic. Significant efforts have been on pillar array surfaces, where the effects of surface chemistry and roughness on contact angle, hysteresis, and wetting stability in terms of the Cassie–Baxter-to-Wenzel transition have been investigated [63–66]. In pillar array surface, roughness and solid-area fraction are controlled by the pillar dimension and geometry, length scale, and pillar height. These surfaces are usually fabricated by lithographic technique or e-beam etching followed by surface hydrophobization with an alkysilane or a perfluoroalkysilane. In 2009, Byun and co-workers reported the preparation of 5×5 µm square pillar (height = 5 µm) array surfaces of varying pitch to model the wettability of insect wings [66]. The model surfaces were prepared by photolithography on silicon wafer and the final textures were surface modified with a fluorosilane layer FOTS, a fluo-rinated   self-assembled-monolayer   synthesized   from   tridecafluoro-1,1,2,2-

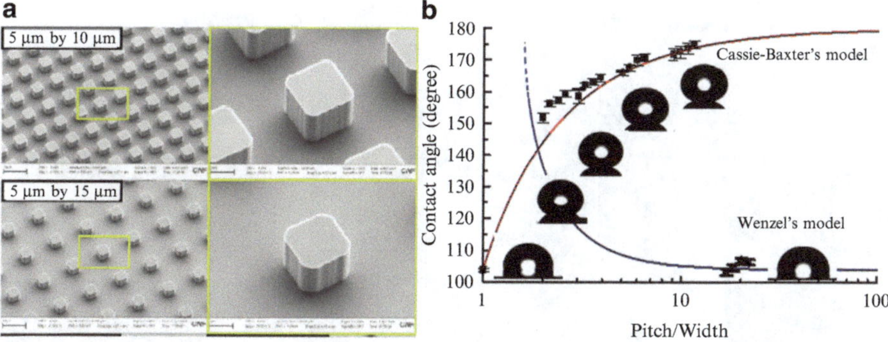

Fig. 4.20 (**a**) Representative SEM micrographs of square pillar array FOTS surfaces and (**b**) plot of water contact angle as a function of pitch/width ratio of the pillar array surface (Reproduced with permission from [66], Copyright 2009 Elsevier)

tetrahydrooctyltrichlorosilane. Representative SEM micrographs of the surfaces are given in Fig. 4.20a. The water contact angles of the model surfaces are plotted as a function of the pitch/width ratio (Fig. 4.20b). Results show that contact angle increase as pitch/width ratio increases initially. At pitch/width ratio between 2 and 15, large contact angles which are predicted by the Cassie–Baxter model are observed, indicating that these surfaces are superhydrophobic. Transition occurs at pitch/width ≈ 15. Water droplets are in the Wenzel state when pitch/width is ≥ 20, indicating that the sagging water interface touches bottom of the pillar array surface. The entire surface becomes fully wetted as a result.

Very similar results were also reported by Oner and McCarthy [63], who found that their 8×8 µm square pillar array surface (height 40 µm) was in the Wenzel state when the pitch is ≥ 56 µm. For 8×8 µm square pillar array with pitches ≤ 32 µm, θ_A is essentially constant at ~174°. On the other hand, θ_R is shown to decrease as the pitch decreases. The result is attributed to the increase in contact line length per unit area. Water pins at the edge of the pillar top, the smaller the pitch, the longer the contact line per unit area, the smaller the θ_R. This interpretation is supported by studying the θ_R values of pillar array surface with different pillar geometry.

For the bottom-up approach, the rough surface can be created by polymerization of a monomer via plasma-enhanced CVD technique or electropolymerization. Surface morphology is sensitive to the polymerization condition. Usually, the roughness is random with a hierarchical particulate or fibrous structure. When a hydrophobic monomer is used, the surface will be superhydrophobic after polymerization. Figure 4.21a shows a 10×10 µm AFM image of a PECVD polymer film polymerized from perfluorooctyl acrylate. The surface comprises nanospherical particles and exhibits very high water repellency with θ_A/θ_R at 168°/165° [67]. Similarly, electropolymerization of 3,4-ethyleneoxythiathiophene derivatives also yield superhydropho-

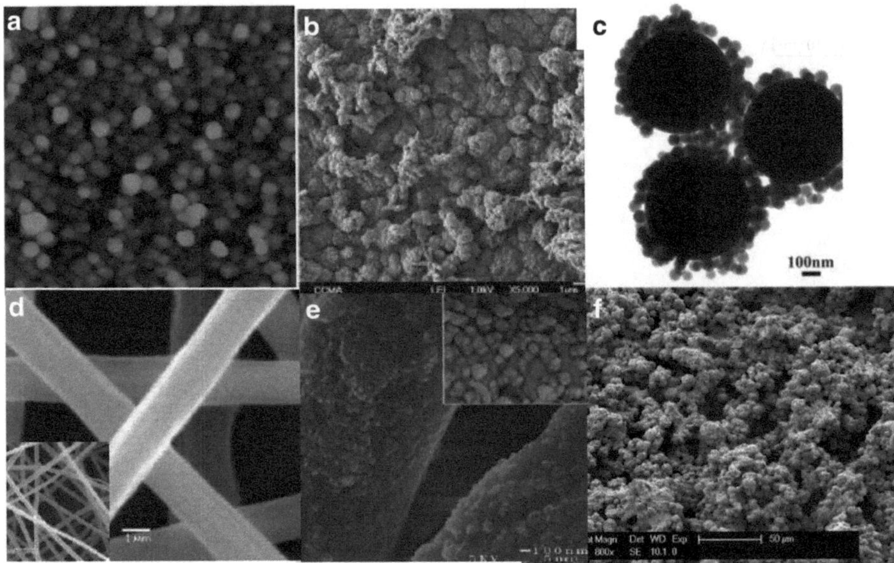

Fig. 4.21 Mosaic of rough superhydrophobic surfaces prepared by different techniques (**a**) PECVD, (**b**) electropolymerization, (**c**) grafting by amine-epoxy reaction, (**d**) electrospinning, (**e**) electrospinning plus layer-by-layer, and (**f**) spray

bic surfaces with θ_A ~160° and low hysteresis. The surface morphology varies from particulate to fibrous and an example is given in Fig. 4.21b [68].

Other popular bottom-up approaches to build rough surfaces include the uses of silica particles or micro/nanofibers from electrospinning [69] or simply spraying a particle-polymer solution onto a substrate. Ming and co-workers [70] reported the preparation of dual scale rough epoxy surface consisting of raspberry-like particles. A schematic for the preparation of the silica-epoxy film is given in Fig. 4.22. First, the raspberry-like particles are prepared by grafting 70 nm amine-functionalized silica particles onto the larger 700 nm epoxy-functionalized silica particles. A TEM micrograph of the raspberry-like silica particles is given in Fig. 4.21c. The superhydrophobic surface is created by grafting the raspberry particles onto an epoxy substrate followed by hydrophobization of the hierarchical surface with a silicone material. The resulting surface was shown to exhibit a θ_A of 165° with a 3° roll-off angle. Ma et al. [71] prepared fiber mat of diameter of ~1.7 μm with average pore size of 80 nm using the electrospinning technique (Fig. 4.21d). After hydrophobization of the fiber mat with a fluoropolymer, the resulting fiber mat is superhydrophobic with θ at ~150°. Fiber mats with hierarchical multi-scale roughness, such as by creating nano pores on the fiber surface or the use of layer-by-layer technique to decorate the electrospun fiber with nano silica particle (Fig. 4.21e), are known to further enhance the water repellency [71]. Spray is a scalable technique that can coat large surface area of different geometry. Steele and co-workers applied the technique to prepare nanocomposite coatings containing 50 nm ZnO nanoparticles

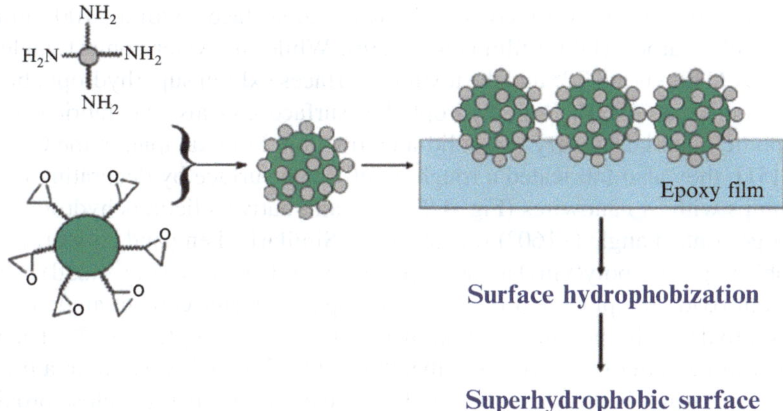

Fig. 4.22 A schematic for the preparation of raspberry-like particles and a superhydrophobic surface with multi-scale roughness (Reproduced with permission from [70], Copyright 2005 The American Chemical Society)

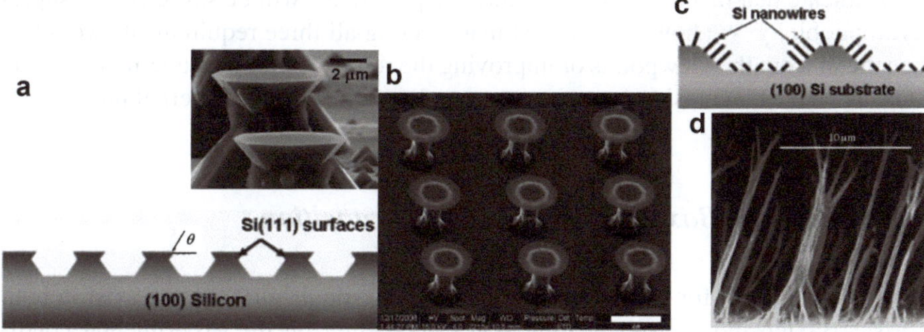

Fig. 4.23 Mosaic of superhydrophobic surfaces prepared with moderately hydrophilic materials/surfaces (**a**) silicon, (**b**) DLC coating (scale bar 20 µm), (**c**) silicon nanowire, and (**d**) poly(vinyl alcohol) nanofibers (scale bar 10 µm)

and a waterborne perfluoroacrylic polymer on a substrate [72]. The resulting surface comprises a hierarchical roughness structure (Fig. 4.21f) and exhibits superhydrophobic property (water $\theta > 150°$ with low hysteresis).

One of the less recognized approaches to design superhydrophobic surface is the use of re-entrant/overhang structure or very low solid-area fraction at the composite interface. Because of the high water surface tension, superhydrophobicity can be attained with moderately hydrophilic materials. Using photolithography, Cao et al. [51] fabricated a pillar array surface with an overhang on silicon wafer (Fig. 4.23a). Since no extra coating is used, the surface of the textured surface is basically hydrogen-terminated silicon which has a contact angle of 74°. The water contact angles of the textured surfaces range between 150 and 160° at solid-area fraction ≤ 0.07. Similarly, Wang et al. [73] created superhydrophobic surfaces with T-shape pillars on silicon wafer followed by coating the textured surfaces with a ~100 nm thick diamond-like-carbon (DLC) film (Fig. 4.23b). While the water contact angle of a smooth DLC film is at ~72° and the textured surfaces exhibit superhydrophobic-like contact angles at ~160°. Superhydrophobic surface can also be fabricated with hydrophilic materials at very low solid-area fraction. For example, in the Cao et al. study [51], they also fabricated a rough, multi-scale surface by decorating the silicon bumps with Si nanowires (Fig. 4.23c). Again, native silicon is hydrophilic but very large contact angle (~160°) was obtained. Similarly, Feng and co-workers [74] were able to grow a poly(vinyl alcohol) (PVA) nanofiber forest (Fig. 4.23d) using an aluminum oxide template and observed an apparent water contact angle of 171°. PVA is a hydrophilic polymer with known water contact angle at ~72°. The commonality of the latter two surfaces is that they all have very low solid-area fraction. Even though water is pinned at the tip of the nanowire or fiber, the close proximity of the pinning location, coupled with the high water surface tension, enables air pockets to be formed and superhydrophobicity. In summary, the design rules for superhydrophobicity are: chemistry, roughness, and re-entrant structure or very low solid-area fraction. The very low solid-area fraction can be achieved using a hierarchical textured approach or a nanowire or nanoforest approach. The above studies also indicate that meeting two out of three requirements will be sufficient for superhydrophobicity. We however suggest that meeting all three requirements would be beneficial from the viewpoints of improving the robustness of the superhydrophobic state as well as increasing its resistance to collapse under high external pressure.

4.3.3 Cassie–Baxter to Wenzel State Transition

It has been known for some time that micro/nano rough surfaces can exist in both Cassie–Baxter and Wenzel wetting states depending on the experimental conditions. For example, He and co-workers [75] showed that water droplet is in the Cassie–Baxter state on the PDMS microtextured surface if the drop is dispensed gently. On the other hand, a Wenzel droplet is obtained when the drop is dispensed at some height. Photographs of these two droplets are reproduced in Fig. 4.24. The results clearly indicate that the Wenzel droplet is the more stable state. In a slightly different experiment, Quere et al. [76, 77] created a Cassie–Baxter droplet with θ_A of 163° on a fluorinated microtextured surface. This Cassie droplet can be converted to the Wenzel droplet when an external pressure is applied. A similar Wenzel droplet can also be created on the same surface by a condensation experiment. While Bormashenko and team [78, 79] showed that the Cassie water droplet can be de-pinned and formed the Wenzel droplet via vibration noise, Boreyko and Chen [53] on the other hand demonstrated that the Wenzel drops obtained by condensing water on the Lotus leaf at high humidity can be converted to the Cassie droplet by vibration too! Another example of "exciting" the Wenzel droplet to the Cassie droplet is provided by Liu and co-workers [80] who were able to turn a Wenzel droplet to

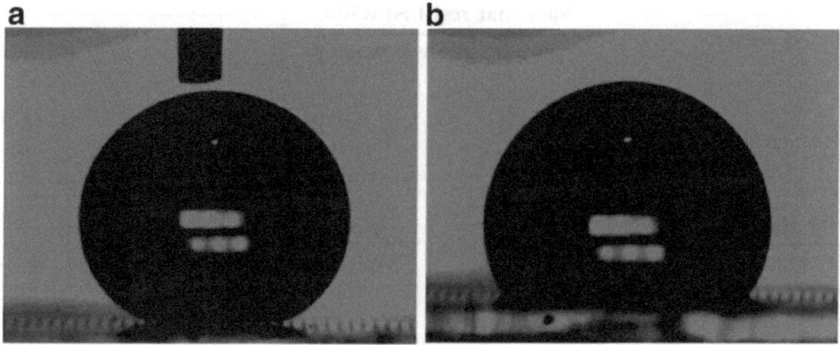

Fig. 4.24 Photographs of (**a**) Cassie and (**b**) Wenzel water droplet obtained on the same microtextured PDMS surface (Reproduced with permission from [75], Copyright 2003 The American Chemical Society)

the Cassie droplet by heating. The overall results seem to suggest that although the Wenzel state may be more stable, the fact that they are interconvertible depending on the experimental condition implies that (1) there exists an energy barrier between them and (2) their difference in energetics is probably small.

In most practical applications, surfaces with high water repellency and self-cleaning property are desired. Approaches to enhance the stability of the Cassie–Baxter state include increase the energy barrier between these states or better yet stabilize and make Cassie–Baxter state the more stable state. Both theoretical analysis and experimental results [81–85] indicate that having a hierarchical, multi-scale roughness structure, a re-entrant structure, along with a hydrophobic surface coating would increase the energy barrier between the Cassie–Baxter and Wenzel state. The increase of barrier height not only enhances the stability of the metastable Cassie–Baxter state, it will also increase its resistance to collapse against external pressure. Moreover, there is preliminary indication that, under certain hierarchical roughness structure, the Cassie–Baxter state may be more favorable thermodynamically [84, 85].

4.3.4 Superoleophobicity

The surface tension of hydrocarbon oil is a lot lower than water. Hexadecane has frequently been used as the probing liquid for oleophobicity, and its surface tension is almost three times lower than that of water (27.5 vs. 72.3 mN/m). It wets most surfaces as a result. Even more challenging is that smooth surface with hexadecane $\theta > 90°$ is not known. Thus, surface with super hexadecane repellency is rare. The consensus definition for superoleophobicity is surface exhibiting hexadecane contact angle >150° and sliding angle ~10°. Unlike superhydrophobic surface, which can be superoleophilic [86, 87], superoleophobic surface is expected to be more versatile as it will repel most liquids ranging from water to alkanes.

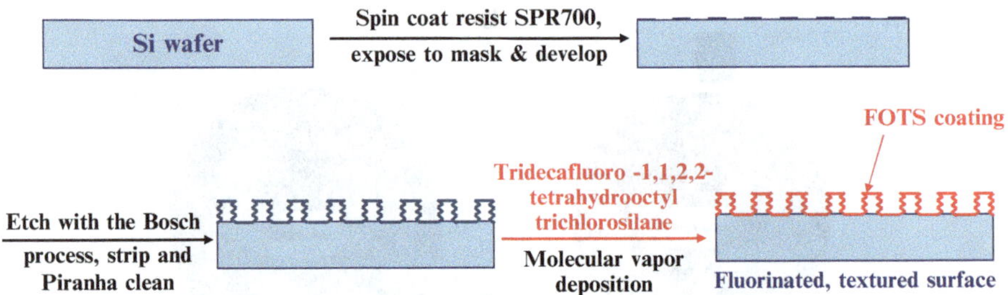

Fig. 4.25 Schematic for the fabrication of fluorinated surface texture on silicon wafer (Reproduced with permission from [88], Copyright 2011 The American Chemical Society)

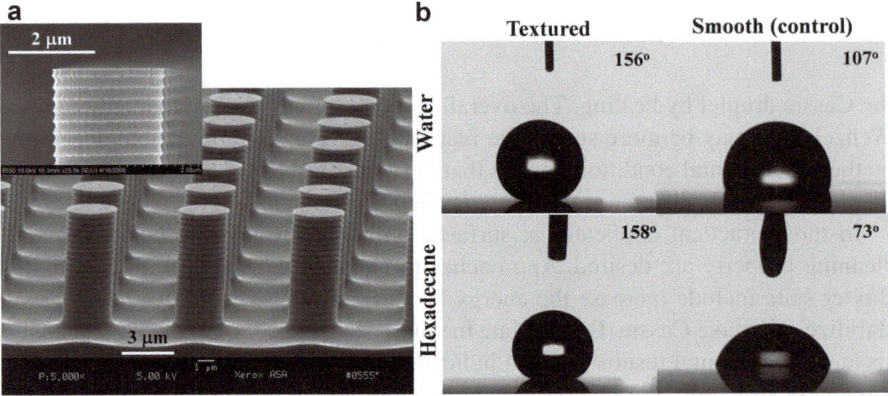

Fig. 4.26 (**a**) SEM micrograph of the textured FOTS surface on Si-wafer (inset: higher magnification micrograph showing details of the pillar structure) and (**b**) static contact angles for water and hexadecane on the textured FOTS surface (control: smooth FOTS surface on Si-wafer) (Reproduced with permission from [88], Copyright 2011 The American Chemical Society)

To elucidate the basic design parameters for superoleophobicity, Zhao et al. [88] launched a systematic investigation using model pillar array surface with well-defined texture and geometry. Figure 4.25 summarizes the surface texturing and chemical modification procedure for the fabrication of the model textured surface. Details of the fabrication processes have been published elsewhere [88].

Figure 4.26a shows the SEM micrograph of the FOTS pillar array surface, comprising ~3 μm diameter pillar arrays ~7.8 μm in height with a pitch of ~6 μm. High magnification SEM micrograph (insert) reveals that the sidewall in each pillar consists of a ~300 nm wavy structure from top to bottom, attributable to the Bosch etching process. The surface property of the FOTS pillar array surface was studied by both static and dynamic contact angle measurement techniques and the static contact angle data with water and hexadecane as test liquids are given in Fig. 4.26b. The water and hexadecane contact angles for the FOTS pillar array surface are at 156° and 158°, respectively, and are significantly higher than those of the controlled smooth surfaces, which are at 107° and 73°, respectively. The results suggest that the high contact angles observed for the FOTS pillar array surface are the result of both surface texturing and fluorination. The sliding angles for the FOTS pillar array surface are found to be ~10° with both water and hexadecane. The high contact angles coupled with the low sliding angles lead to the conclusion that the FOTS pillar array surface is both superhydrophobic and superoleophobic with low hysteresis.

The use of photolithography and hydrophobic coatings to create micro/nano textured surfaces is not new [63–66]. The sidewalls of all the reported pillar structures were smooth, and only superhydrophobicity was reported, not superoleophobicity. Recently, Tuteja and co-workers [89] reported the fabrication of electrospun mats that exhibited superoleophobicity. The mat comprises nanofibers made from F-POSS (*1H,1H,2H,2H*-heptadecafluorodecyl polyhedral oligomeric silsesquioxane) and PMMA (poly(methyl methacrylate)) blends. The flat surface of the same material is oleophilic with a hexadecane advancing contact angle of ~80°. To elucidate the mechanism for the superoleophobicity of the electrospun mat, these authors created the so-called micro hoodoo structure on silicon wafer using photolithography and surface fluorination. These authors concluded that the re-entrant geometry in the "micro-hoodoo" structure is critical to achieving superoleophobicity and the electrospun mat has a similar geometry at the liquid–solid–air composite interface. The conclusion is substantiated by additional experimental and modeling studies [86, 87, 90].

To elucidate the origin for the observed superoleophobicity, FOTS pillar array surfaces with (a) a smooth straight sidewall and (b) a straight sidewall with an overhang structure were fabricated (Fig. 4.27). The wetting properties with water and hexadecane were examined and key sessile drop data are included in the figure. The

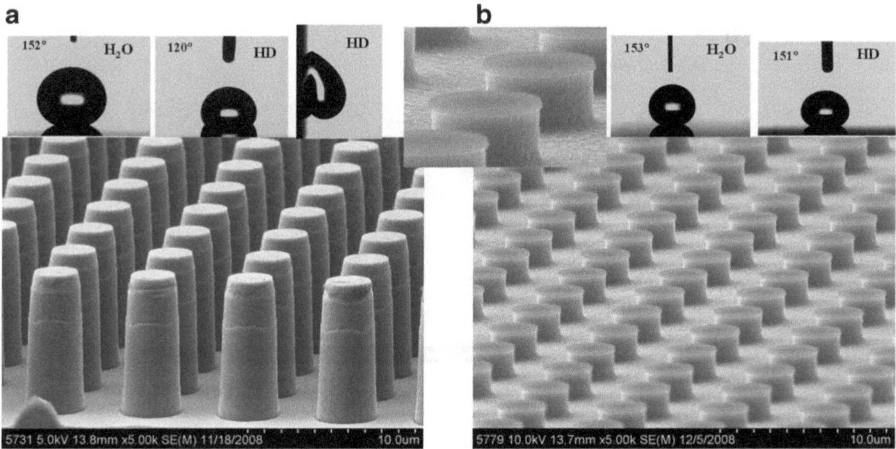

Fig. 4.27 (**a**) SEM micrograph of the textured FOTS surface with straight smooth sidewall pillars on Si-wafer; and (**b**) SEM micrograph of the textured FOTS surface with an overhang pillar structure on Si-wafer (insert: sessile drop data for water and hexadecane on the textured surfaces) (Reproduced with permission from [88], Copyright 2011 The American Chemical Society)

results show that both surfaces are superhydrophobic, but that there is a significant difference in oleophobicity. For pillar array surface with a smooth, straight sidewall, the hexadecane drop exhibits a hexadecane contact angle of 121°, but the drop was found stuck on the surface when tilt to 90°. On the other hand, the surface with an overhang pillar was shown to be superoleophobic with hexadecane contact angle and sliding angle at 150° and 12°, respectively. The similarity in wetting properties between the model surface with the wavy sidewall and the pillar surface with the overhang structure leads to the conclusion that the re-entrant structure at the top of the wavy sidewall is a key contributor to the superoleophobic property. This conclusion is not only in agreement with those reported by Tuteja and co-workers [86, 87, 89, 90], but also consistent with other observations reported by Fujii et al. [91], Cao et al. [92], Ahuja et al. [93], and Kumar and co-workers [94]. They all pointed to the importance of the re-entrant or overhang structure in achieving surface superoleophobicity. Hence, the basic parameters for superoleophobicity are: chemistry (contact angle of the coating), roughness, and re-entrant geometry. Unlike superhydrophobicity where meeting two out of three requirements is sufficient, superoleophobicity demands all three requirements attributable to the lowering of the surface tension of the probing liquid.

An attempt was made to visualize the composite interface using hot polyethylene wax. Figure 4.28a shows the image of a hot polyethylene wax sessile droplet on the model pillar array surface at 110°C. The contact angle and the sliding angle of the droplet were measured at 155° and 33°, respectively, consistent with the hexadecane data. The data suggests that the wax droplet is in the Cassie–Baxter state on the pillar array surface (Fig. 4.28b). When the wax droplet was cooled to room temperature, it was carefully detached from the textured surface and the geometry of the composite interface was examined by SEM microscopy. Figure 4.28c shows the

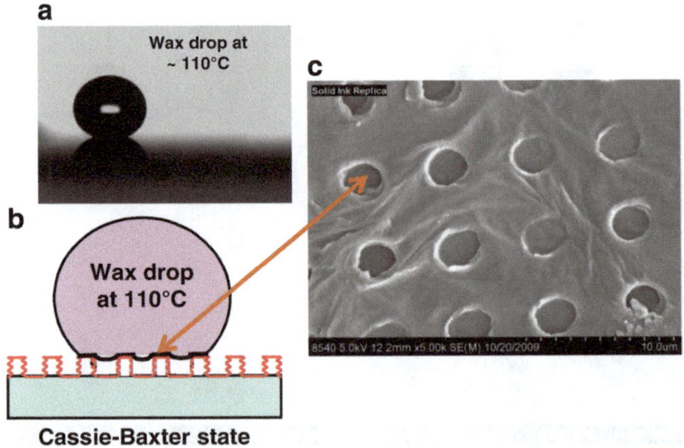

Fig. 4.28 (**a**) Molten droplet of polyethylene wax on the pillar array surface, (**b**) wax droplet in the Cassie–Baxter sate, and (**c**) SEM micrograph of the wax replica of the composite interface after the wax droplet is cooled to room temperature (Reproduced with permission from [88], Copyright 2011 The American Chemical Society)

SEM micrograph of the liquid–solid–air composite interface. The result indicates that the wax surface is "flat" with holes corresponding to the location of the pillars. From the thickness of the "rim," one can estimate the penetration depth of the molten wax droplet into the void space between the pillars, which is <0.5 μm. Since the height of the pillar is ~7 μm, this result positively reveals that the molten wax droplet is indeed sitting on air at the interface of the superoleophobic surface. However, the composite interface is not perfectly flat; the molten wax appears to penetrate into the void space between the pillars. Although the composite interface for water on the FOTS pillar array surface was not imaged, we believe that the water droplet is in the Cassie–Baxter state too based on the contact angle and sliding angle data.

The effect of surface texturing, including pillar pitch, diameter and height, on both static and dynamic contact angles have also been studied [95]. The contact angle results (θ_A, θ_R, α, and hysteresis) are plotted as a function of solid-area fraction and are summarized in Fig. 4.29. The results show that surface texturing has very little effect on the static and advancing contact angles for both water and hexadecane. On the other hand, pillar spacing has profound effects on the dynamic contact angles, particularly the sliding angle, the receding contact angle and the contact angle hysteresis. θ_R decreases, and α and ($\theta_A-\theta_R$) increase for both water and hexadecane, indicating that surface adhesion increases and drop mobility decreases as the solid-area fraction increases. This is attributed to the pinning of the liquid droplet on the pillars, the larger the solid-area fraction the more the pinning sites (or the higher the contact line density). It is important to point out that the effect of solid-area fraction is larger for hexadecane, and this is due to their difference in the pinning location [95].

Also included in Fig. 4.29a, b are results from pillar array surfaces with 1 and 5 μm diameter pillars (represented by data points X and O, respectively). These data points are completely compatible with the results of the 3 μm pillar array surfaces, indicating that surface adhesion and drop mobility are governed primarily by the density of the contact lines, not the geometry of the texture.

The effect of pillar height on the superoleophobic properties was studied using 3 μm pillar array surfaces with a center-to-center spacing of 12 μm. This pillar spacing is wider than the surface shown in Fig. 4.26 and represents a more stressful case for liquid sagging. The pillar height was controlled by the number of Bosch-etching cycles and is varied from about ~0.8 to 8 μm. Figure 4.30 shows the SEM micrographs of three representative surfaces along with the plots of contact angle data against the pillar height for both water and hexadecane. The results show that pillar array surfaces can maintain its Cassie–Baxter state with both water and hexadecane as long as the pillar height is >1 μm. For surfaces with pillar height ≤0.8 μm, both water and hexadecane droplets are unstable. They are shown to wet the entire surface sooner or later. This indicates that the liquid droplets sag and touch the bottom of the surface fully wetting it as a result. To avoid wetting transition from the Cassie–Baxter to Wenzel state, pillar height should be >1 μm.

In summary, we show with model pillar array surfaces that hydrocarbon oils can trap air pockets on rough surfaces and result in a Cassie–Baxter composite interface when the following basic parameters are met: high hexadecane contact angle surface

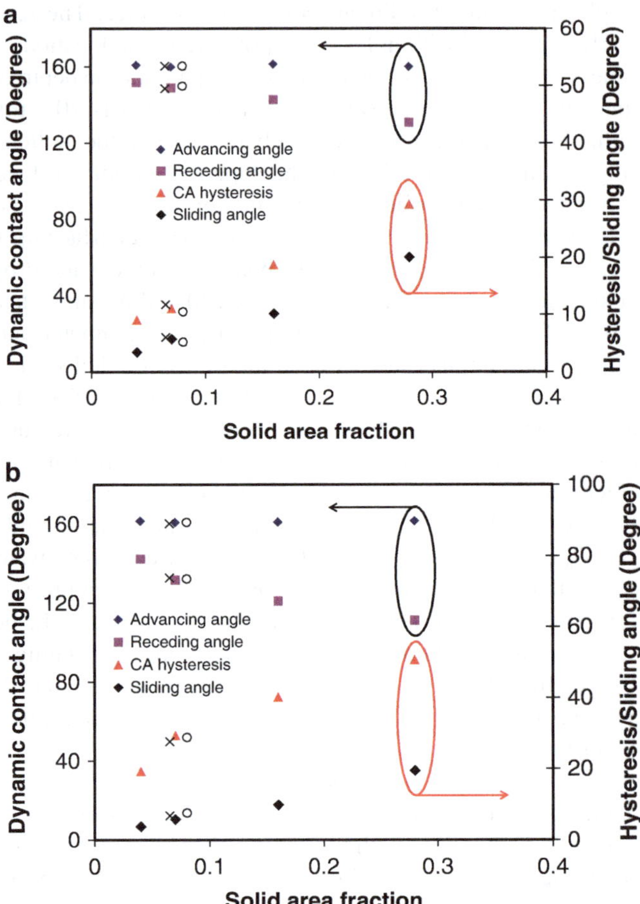

Fig. 4.29 Plots of dynamic contact angle data with (**a**) water and (**b**) hexadecane versus solid-area fraction for 3 μm pillar array surfaces. (*Insert*: dynamic contact angle data from a 1 μm pillar array (X) and a 5 μm pillar array (O) surface) (Modified from figures in [95], Copyright 2012 The American Chemical Society)

coating and a rough surface with an overhang or re-entrant structure. Since oleophobic coating does not exist, the re-entrant structures, which allow the hydrocarbon liquid to pin at the composite interface is essential in achieving superoleophobicity.

4.4 Directional Wetting and Spreading

In most rough surfaces, whether the texture is patterned regularly or randomly, wetting and spreading on them are isotropic two dimensionally. However, if the texture or roughness is created with different wettability directionally, anisotropic wetting and spreading will occur. This happens in nature as well as artificial surfaces. Two famous examples in nature are the groove structures in the wings of butterfly [96] and the directional papillae array structure in rice leaf [97]. Figure 4.31a depicts a picture of an iridescent blue butterfly. The SEM micrograph of the wing (Fig. 4.31b) shows that the wing comprises layers of "leaves" stacking outward

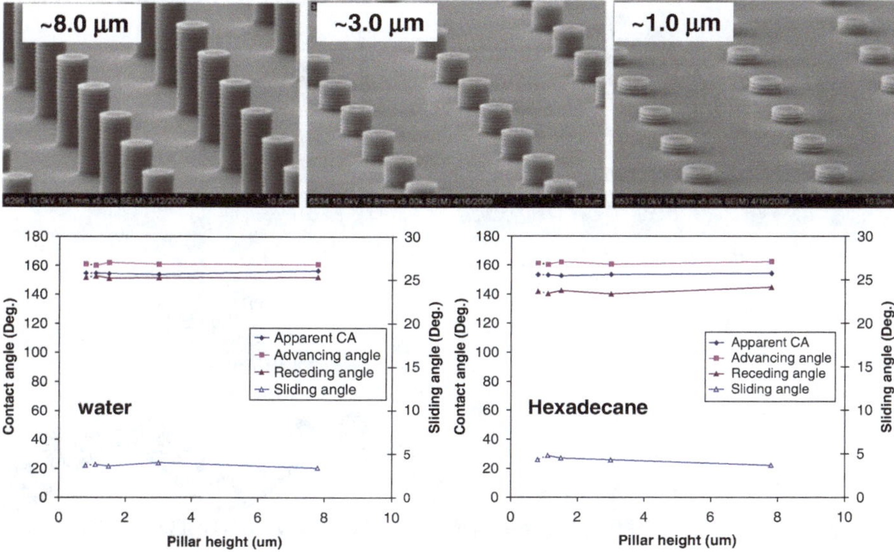

Fig. 4.30 (*top*) SEM micrographs of 3 μm pillar array surfaces with different height and (*bottom*) plots of static and dynamic contact angles of the 3 μm pillar array surfaces as a function of pillar height

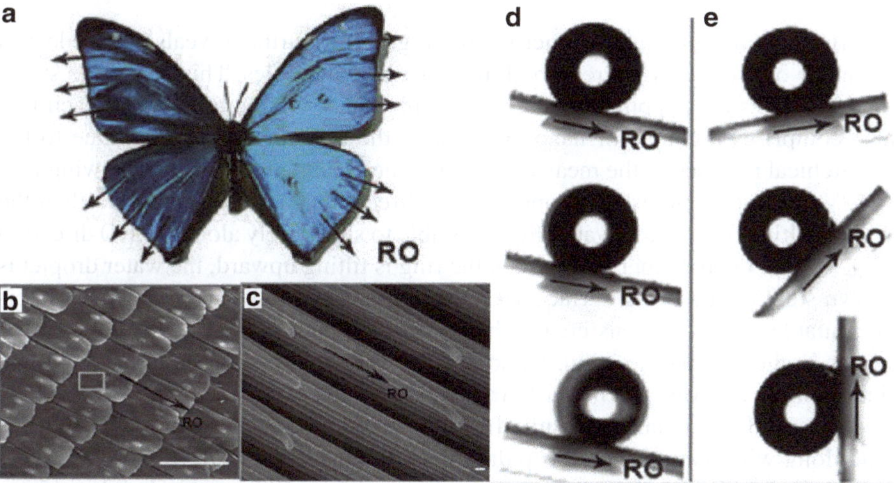

Fig. 4.31 (**a**) Photograph of an iridescent blue butterfly, (**b**) SEM micrograph of the wing of butterfly, scale bar 100 μm, (**c**) high magnification SEM micrograph of the wing of butterfly, scale bar 100 nm, and (**d, e**) water sessile drop data in outward and inward direction, respectively (Reproduced with permission from Ref. 96, Copyright 2007 The Royal Society of Chemistry)

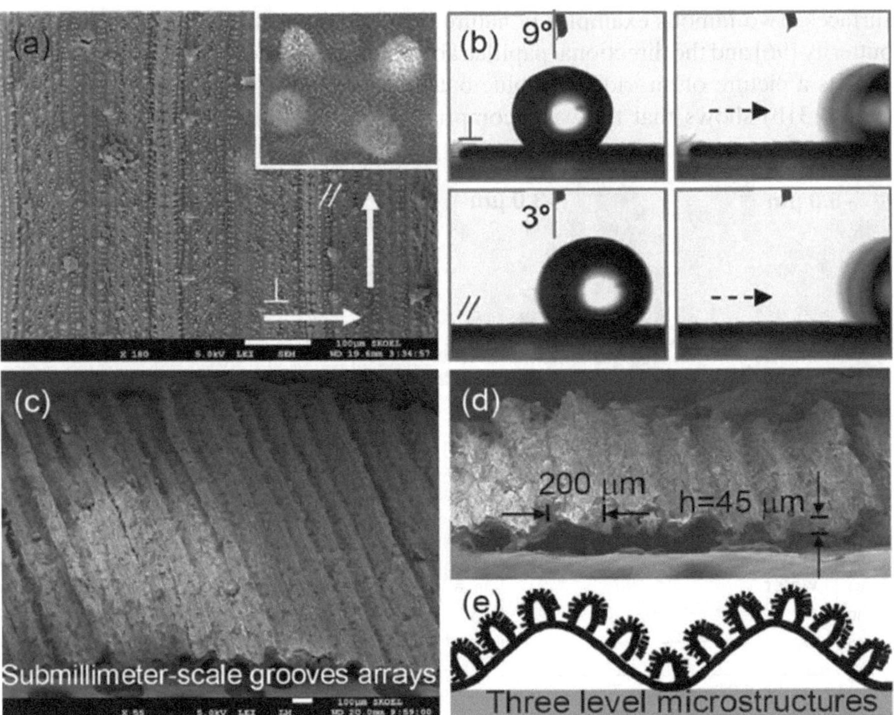

Fig. 4.32 (**a**) SEM micrograph of a rice leaf, scale bar 100 µm, (inset: highly magnified SEM), (**b**) water sessile drop data on the leaf surface, (**c**) SEM micrograph at 60° tilt angle, scale bar 100 µm, (**d**) cross-section SEM micrograph, and (**e**) a schematic of the hierarchical surface structure of the rice leaf. (Reproduced with permission from Ref. 97, Copyright 2011 Wiley)

orderly. High resolution SEM micrograph (Fig. 4.31c) further reveals that the leaves are made of groove structure in both micron and nanoscales. This dual scale structure facilitates air-trapping when contact with water. On the top of that, each leaf also comprises a couple of nano hooks with the tips curving inward. Due to the hierarchical roughness, the measured water contact angle for the butterfly wing is at ~152°. The most interesting property is its directional wetting property. When the wing is tilting downward, water droplet is able to slide freely along the RO direction (Fig. 4.31d). On the other hand, when the ring is tilting upward, the water droplet is shown stuck on the wing surface even when it is tilted to 90° (Fig. 4.31e). This is attributable to the frictions created by the nano hooks as well as the edge of the leaves in the against direction. Undoubtedly, after millions of years of evolution, this is one of the ways the butterfly tries to keep itself dry in the rain.

Figure 4.32 summarizes results from the SEM microscopy studies of the rice leaf along with water sessile drop data [97]. In terms of surface texture, rice leaf exhibits submillimeter groove structure with width at ~200 µm and height ~45 µm (Fig. 4.32c, d). Each groove is made of micron-size papillae array lining up in the direction of the groove, and the entire surface is covered with nano-size hairy plant wax (Fig. 4.32a). This hierarchical surface structure has resulted in a very large water contact angle (~150°). Due to directionality of the papillae arrangement, water droplet displays anisotropic sliding angle. Specifically, the sliding angles were found to be 3° and 9° in the parallel and perpendicular direction, respectively. As a surface design, the superhydrophobic nature of rice leaf leads to self-cleaning, ensuring efficient absorption of the sunlight for photosynthesis during the growing season. At the same time, the directional water sliding will make sure water and nutrient are delivering to the root effectively, again for growth and survival.

Mimicking nature, many artificial groove structured surfaces ranging from micron to nano length scale of varying hydrophobicity have been reported [98–101]. Typically, the droplet is elongated. The distortion is a function of the geometry of the groove (width and spacing), the drop size, and the wettability of the surface. Figure 4.33a shows a schematic for a series of 3 µm deep, PMMA groove surfaces with width (spacing) varying from 5, 10, 25, 50, to 100 µm. These surfaces were made by the hot embossing technique with appropriate lithographically prepared silicon molds [101]. After molding, the surfaces were modified by a plasma polymerized coating. The wettability of the surface coating is controlled by the type of monomer used. Two monomers, namely allylamine and hexane were used and they provide a hydrophilic

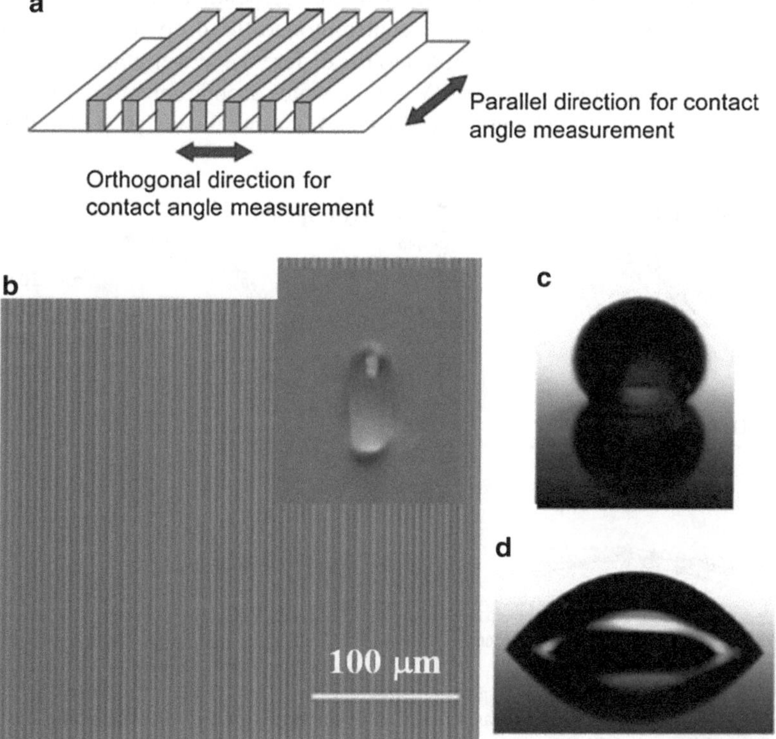

Fig. 4.33 (**a**) A schematic of the groove PMMA surface, (**b**) optical image of a 5 µm groove surface, and (**c**, **d**) sessile drop data on the surface in the perpendicular and parallel direction (Reproduced with permission from [101], Copyright 2009 The American Chemical Society)

and hydrophobic surface coating with water contact angles at ~60° and ~98°, respectively. Figure 4.33b shows a photograph of a typical surface and the shape of the water sessile droplet is given in the insert. Figure 4.33c, d show the water sessile drop data in the perpendicular and parallel direction, respectively. It is clear that the water droplet is elongated in the parallel direction due to wetting interaction between water and the groove surface. The contact angle in the perpendicular direction is always larger than that in the parallel direction.

Yang et al. [101] went on to investigate the effects of groove width, drop size, and coating chemistry on the wettability of the groove surfaces. The water contact angle data, by plotting the contact angles as a function of the chemistry in the perpendicular and parallel direction, are summarized in Fig. 4.34a, b. The upper and lower dash lines in the plots are the calculated Cassie–Baxter and Wenzel angles, respectively, based on the surface chemistry and texture geometry. Not unexpectedly, there is little agreement between the calculated Cassie–Baxter/Wenzel angles with the experimental values. For groove surfaces with the hydrophilic coating, they are expected to be in the fully wetted by water. The results in Fig. 4.34a clearly show that (1) the parallel direction is more wettable (smaller contact angle) than the perpendicular direction, and (2) the distortion of the drop increases as the size of the water droplet increases. While increase of drop size has little effect on the contact angle in the parallel direction, the effect in the perpendicular direction is much larger. In fact, this effect is the prime contributor to the drop size effect on drop elongation. The observation is attributed to the difference in wettability between the parallel and perpendicular direction. When the water drop is spreading in the parallel direction, there is little change in solid-area fraction, always 50 %. On the other

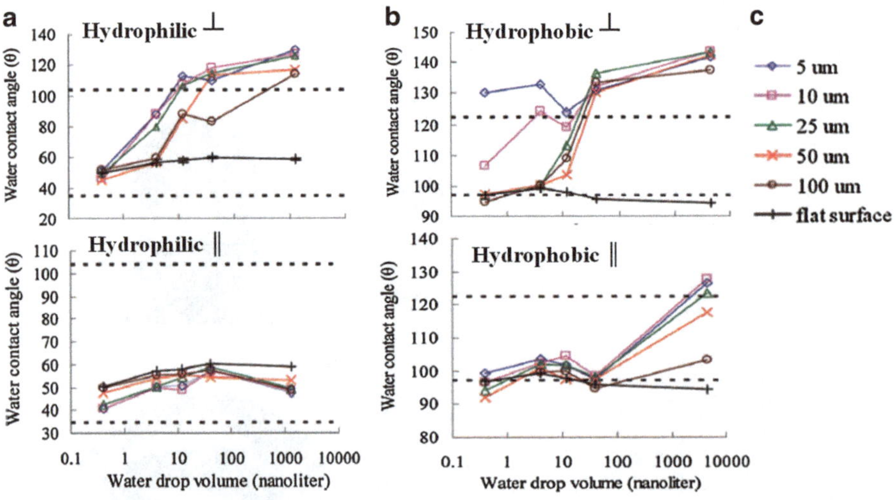

Fig. 4.34 Plot of water contact data as a function of the drop size for (a) hydrophilic coated and (b) hydrophobic coated PMMA groove surfaces (*top*: measured in the perpendicular direction, *bottom*: measured in the parallel direction), and (c) width of the grooves in (a) and (b) (Modified from [101], Copyright 2009 The American Chemical Society)

hand, as the contact line is advancing perpendicularly, it is experiencing friction due to pinning at the edge of the groove 50 % of the time, the larger the drop size, the more friction the drop experiences during advancing. The largest retardation to contact line advance is expected for groove surface with the narrowest width (5 μm). Indeed, the largest water contact angle is observed for the 5 μm groove surface in the perpendicular direction.

With the hydrophobic coating (Fig. 4.34b), the groove surface can be in the Wenzel or Cassie–Baxter state depending on the drop size. Although anisotropic wettability is observed, there is no clear trend between drop elongation and drop size as the case for the hydrophilic groove surfaces. From the size of the contact angle, it appears that water droplet is in the Cassie–Baxter state when the droplet is >50 nL. The droplet appears to be in the Wenzel state for the 0.5 nL drop and then transitioning to the Cassie–Baxter state as the drop size increases.

Directional self-cleaning superoleophobic groove surface was reported by Zhao and Law in 2012 [102]. The surface, prepared by photolithographic technique followed by surface fluorination, comprises 4 μm deep, 3 μm wide groove structure with a 6 μm pitch. Figure 4.35a shows the SEM micrograph of the groove surface. Due to the Bosch etching process, wavy sidewall similar to those in the pillar array surface was obtained. The water and hexadecane sessile drop data are summarized in Fig. 4.35b, and anisotropic wettability is evident for both liquids. The contact angles in the orthogonal direction for water and hexadecane are at 154° and 162°, respectively, and are consistently larger than those in the parallel direction, which are at 131° and 113°, respectively. The large contact angles suggest that both liquids are in the Cassie–Baxter state on the groove surface. Interestingly, the sliding angles for water and hexadecane are found to be smaller in the parallel direction, 8° and 4° as compared to 23° and 34° in the orthogonal direction, despite having smaller contact angles. The overall wetting data can be rationalized based on the pinning effect. For example, due to the friction created at the edge of the sidewall, liquid advance in the orthogonal direction will be retarded every time the liquid advances

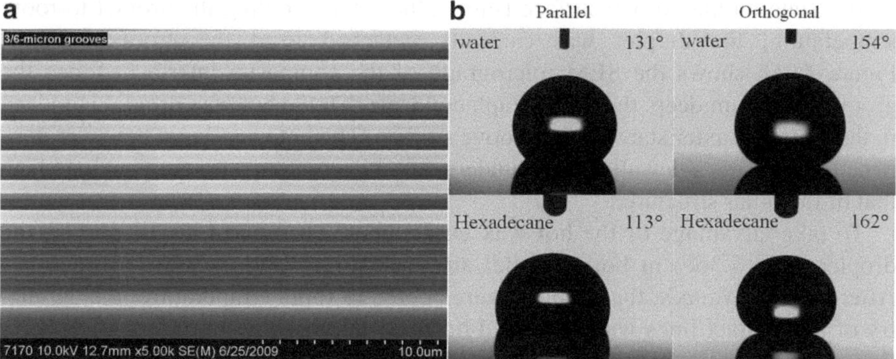

Fig. 4.35 (**a**) SEM micrographs of the groove FOTS surface and (**b**) sessile drop data of water and hexadecane on the groove FOTS surface (Reproduced with permission from [102], Copyright 2012 The American Chemical Society)

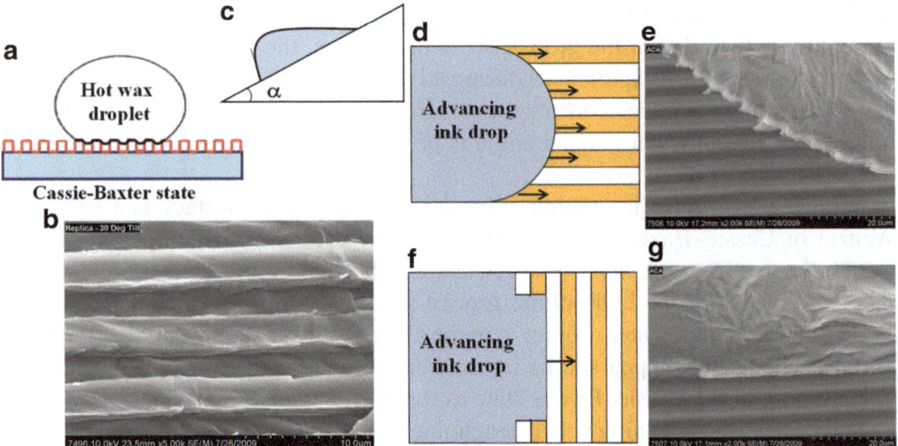

Fig. 4.36 (**a**) A schematic showing the composite interface of the hot wax drop on the groove surface, (**b**) SEM micrograph of the wax replica of the interface, (**c**) schematic of a hot wax droplet on a tilting plate, (**d, e**) schematic and SEM micrograph of the wax drop sliding in the parallel direction, and (**f, g**) schematic and SEM micrograph of the wax drop sliding in the orthogonal direction (Modified from figures in [102], Copyright 2012 The American Chemical Society)

to the next groove. This results in larger contact angle and sliding angle. In contrast, the same friction does not exist when the liquid advances in the parallel direction. The liquid simply wets the solid strip and results in smaller sliding angle and smaller contact angle. This interpretation is supported by the image of the contact lines shown and discussed below.

Similar to the pillar array surface, the composite interface between the groove surface and hydrocarbon liquid was imaged with a hot polyethylene wax droplet. The contact angles for the wax drops at 105 °C are at 156° and 120° in the orthogonal and parallel direction, respectively, indicating that the droplet is in the Cassie–Baxter state on the groove surface (Fig. 4.36a). After cooling the droplet to room temperature, the droplet was carefully detached from the groove surface. Figure 4.36b shows the SEM micrograph of the composite interface. Since the grooves are 4 μm deep, the micrograph confirms that the hot wax droplet is indeed in the Cassie–Baxter state on the groove surface. The hot wax liquid however does penetrate into the sidewall and pin underneath the re-entrant structure analogous to that of the pillar structure.

To take advantage of the hot wax experiment, Zhao and Law tilt the sessile droplet (Fig. 4.36c) in both parallel and orthogonal (Fig. 4.36d, f) directions. After liquid advances, the droplets were cooled to room temperature; the geometry of the contact lines was examined by SEM microscopy. In the parallel direction, the data show that the contact lines advance through wetting the solid strips of the groove surface, resulting in a zig-zag pattern along the contact line (Fig. 4.36e). In the orthogonal direction, liquid advance requires the contact line to "jump" from one solid strip to the next (Fig. 4.36g). Due to the high energy

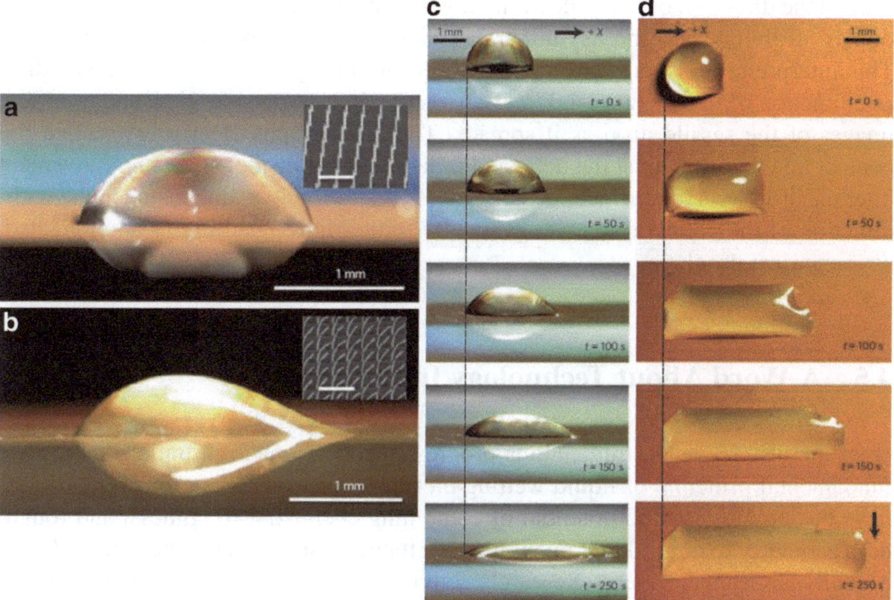

Fig. 4.37 Comparison of wetting behavior of water sessile droplets (0.002 % Triton X-100) on asymmetric nanopillar arrays with (**a**) zero and (**b**) 12° deflection [inset: SEM images of the surfaces, 10 μm scale bar]; (**c, d**) are time-lapse side-view and top-view images showing the spreading of a 1 μL sessile droplet on surface in b (Reproduced with permission from [103], Copyright 2010 Nature Publishing Group)

barrier in this process, the sliding angle in the perpendicular direction is larger and the contact angle is larger too due to the pinning effect.

While having a directional wetting surface is certainly interesting and may find many applications, Chu, Xiao, and Wang [103] reported recently that with a proper structural design, wetting and spreading can be unidirectional too! Using photolithographic technique, these authors first fabricated a pillar array surface with diameter ~500–750 nm, height ~10 μm, and spacing ~3.5 μm. The surface of the pillar array can be modified by a polymer coating using the CVD technique and a typical SEM image of the surface is shown in the inset of Fig. 4.37a. Pillar array surfaces with different degree of pillar deflections (from the vertical position) can be fabricated by first asymmetrically depositing a thin layer of gold on one side of the pillar using the e-beam technique, followed by coating the entire pillar structure with a CVD polymer coating. The angle of deflection is controlled by the thickness of the gold layer. Deflection angle ranges from 7° to 52° were reported and the SEM image of the surface with 12° deflection is shown in the inset of Fig. 4.37b. The water sessile drop data show that with the vertical pillar array surface, the droplet is symmetrical (Fig. 4.37a), whereas asymmetrical drop shape is observed for the surface with a 12° deflection pillar array (Fig. 4.37b). More importantly, a smaller contact angle is obtained when the advancing contact line is moving in the "with-direction" of the deflection. When the contact line is advancing against the direction of the deflection, larger contact angle is resulted. The observation is attributable to the asymmetrical friction created by the deflected pillars. This interpretation is supported when observing the spreading of the sessile drop as a function of time. Figure 4.37c, d shows the time-lapse side-view and top-view images of the sessile drop as it spreads. The photographs clearly reveal that the contact line that is against the deflected pillars (left hand side) shows no movement after the drop wets the surface. On the other hand, the contact line that is in the with-direction of the deflection moves gradually to the right, resulting in an elongated, asymmetrical droplet.

4.5 A Word About Technology Implementation

Over the last 20 years, tremendous progress has been made in understanding the fundamental principle of liquid wetting on rough surfaces. Numerous applications have been explored. The potential of combining chemistry, roughness, and roughness geometry to control and manipulate wettability, spreading, adhesion, and drop mobility appears limitless. Research activities have been diverse and rigorous. While superhydrophilicity did find applications in antifogging coating and self-cleaning surface [35], the overall implementation of the rough surface-enabled technology is still lagged when compared to the worldwide effort and investment. This is particularly true for superhydrophobicity. Many researchers have been inspired by the Lotus effect and the phenomenon seems show promises in a variety of areas. Unfortunately, most of the excitements are just hypes or simply "You Tube" moments. Large-scale adoption of any technology derived from superhydrophobicity remains elusive. There are multiple causes for this shortfall. Part of it can be attributed to the insufficient recognition of all the key parameters needed for production. Most researchers are only focused on the large contact angle and small sliding angle or low hysteresis in their technology pursue. These parameters are crucial, but they are just a starting point. Insufficient attention has been paid to key parameters such as mechanical property, oleophobicity, wetting breakthrough pressure, and manufacturability. These key parameters are essential as they are intimately tied to robustness of the product design and longevity in performance. A lot of superhydrophobic surfaces, which comprise nanoscale fine structures, are fragile mechanically. Additionally, in urban and industrial environments, exposure to volatile organics is unavoidable. If a superhydrophobic surface is contaminated by oily matter, its performance will degrade. Therefore, abrasion resistance and oleophobicity have to be considered seriously in a robust superhydrophobic product design. While the requirement for a high wetting breakthrough pressure may be a dependent of the application, manufacturing technology for large-scale and large area surface manufacturing is a must-have if mass adoption is desired. To date, very little is known about robust design parameters for manufacturing latitude and defect rate. A detailed discussion of these issues is beyond the scope of this book, but can be found in a recent chapter authored by the present authors in an upcoming book titled "Non-wettable Surfaces" edited by Ras and Marmur [104].

References

1. Wenzel RN (1936) Resistance of solid surfaces to wetting by water. Ind Eng Chem 28:988–994
2. Cassie ABD, Baxter S (1944) Wettability of porous surfaces. Trans Faraday Soc 40:546–551
3. Gao L, McCarthy TJ (2007) How Wenzel and Cassie were wrong. Langmuir 23:3762–3765
4. McHale G (2007) Cassie and Wenzel: were they really so wrong? Langmuir 23:8200–8205
5. Nosonovsky M (2007) On the range of applicability of the Wenzel and Cassie equations. Langmuir 23:9919–9920
6. Panchagnula MV, Vedantam S (2007) Comment on how Wenzel and Cassie were wrong by Gao and McCarthy. Langmuir 23:13242
7. Gao L, McCarthy TJ (2007) Reply to "Comment on how Wenzel and Cassie were wrong by Gao and McCarthy". Langmuir 23:13243
8. Gao L, McCarthy TJ (2009) An attempt to correct the faulty intuition perpetuated by the Wenzel and Cassie "Laws". Langmuir 25:7249–7255
9. Pease DC (1945) The significance of the contact angle in relation to the solid surface. J Phys Chem 49:107–110
10. Bartell FE, Shepard JW (1953) Surface roughness as related to hysteresis of contact angles. II. The systems of paraffin-3 molar calcium chloride solution-air and paraffin-glycerol-air. J Phys Chem 57:455–458
11. Johnson RE (1959) Conflicts between Gibbsian thermodynamics and recent treatments of interfacial energies in solid-liquid-vapor systems. J Phys Chem 63:1655–1658
12. Gray VR (1965) Surface aspects of wetting and adhesion. Chem. Ind. 969–977
13. Extrand CW (2003) Contact angles and hysteresis on surfaces with chemical heterogeneous islands. Langmuir 19:3793–3796
14. Bormashenko EA (2009) Variation approach of wetting of composite surfaces: is wetting of composite surfaces a one-dimensional or two-dimensional phenomenon? Langmuir 25:10451–10454
15. Gao L, McCarthy TJ (2009) Wetting 101°. Langmuir 25:14105–14115
16. Meiron TS, Marmur A, Saguy IS (2004) Contact angle measurement on rough surfaces. J Colloid Interface Sci 274:637–644
17. Forsberg PSH, Priest C, Brinkmann M, Sedev R, Ralston J (2010) Contact line pinning on microstructured surfaces for liquids in the Wenzel state. Langmuir 26:860–865
18. Choi W, Tuteja A, Mabry JM, Cohen RE, McKinley GH (2009) A modified Cassie-Baxter relationship to explain contact angle hysteresis and anisotropy on non-wetting surfaces. J Colloid Interface Sci 339:208–216
19. Kanungo M, Mettu S, Law KY, Daniel S (2014) Effect of roughness geometry on wetting and dewetting of rough PDMS surfaces. Langmuir 30:7358–7368
20. Drelich J, Chibowski E (2010) Superhydrophilic and superwetting surfaces: definition and mechanisms of control. Langmuir 26:18621–18623
21. Priest C, Albrecht TWJ, Sedev R, Ralston J (2009) Asymmetric wetting hysteresis on hydrophobic microstructured surfaces. Langmuir 25:5655–5660
22. Papadopoulos P, Deng X, Mammen L, Drotlef DM, Battagliarin G, Li C, Mullen K, Landfester K, del Campo A, Butt HJ, Vollmer D (2012) Wetting on the microscale: shape of a liquid drop on a microstructured surface at different length scales. Langmuir 28:8392–8398
23. Dorrer C, Ruhe J (2006) Advancing and receding motion of droplets on ultrahydrophobic post surfaces. Langmuir 22:7652–7657
24. Hansson PM, Hormozan Y, Brandner BD, Linnros J, Claesson PM, Swerin A, Schoelkopf J, Gane PAC, Thormann E (2012) Effect of surface depressions on wetting and interactions between hydrophobic pore array surfaces. Langmuir 28:11121–11130
25. Anantharaju N, Panchagnula MV, Vedantam S, Neti S, Tatic-Lucic S (2007) Effect of three-phase contact line topology on dynamic contact angles on heterogeneous surfaces. Langmuir 23:11673–11676

26. Liu J, Mei Y, Xia R (2011) A New wetting mechanism based upon triple contact line pinning. Langmuir 27:196–200
27. Larsen ST, Taboryski RA (2009) Cassie-like Law using triple phase boundary line fractions for faceted droplets on chemically heterogeneous surfaces. Langmuir 25:1282–1284
28. Jopp J, Grll H, Yerushalmi-Rozen R (2004) Wetting behavior of water droplets on hydrophobic microtextures of comparable size. Langmuir 20:10015–10019
29. Mohammadi R, Wassink J, Amirfazli A (2004) Effect of surfactants on wetting of superhydrophobic surfaces. Langmuir 20:9657–9662
30. Wu J, Zhang M, Wang X, Li S, Wen W (2011) A simple approach for local contact angle determination on a heterogeneous surface. Langmuir 27:5705–5708
31. Cwikel D, Zhao Q, Liu C, Su X, Marmur A (2010) Comparing contact angle measurements and surface tension assessments of solid surfaces. Langmuir 26:15289–15294
32. Koch K, Bhushan B, Barthlott W (2008) Diversity of structure, morphology, and wetting of plant surfaces. Soft Matter 4:1943–1963
33. Koch K, Bhushan B, Barthlott W (2009) Multifunctional surface structure of plants. An inspiration for biomimetics. Prog Mater Sci 54:137–178
34. Hasimoto K, Irie H, Fujishima A (2005) TiO$_2$ photocatalysis: a historical overview and future prospect. Jpn J Appl Phys 44:8269–8285
35. Fujishima A, Zhang X, Tryk DA (2008) TiO$_2$ photocatalysis and related surface phenomena. Surf Sci Rep 63:515–582
36. Ogawa T, Murata N, Yamazaki S (2003) Development of antifogging mirror coated with SiO$_2$-ZrO$_2$-colloidal SiO$_2$ film by the Sol Gel process. J Sol-Gel Technol 27:237–238
37. Cebeci FC, Wu Z, Zhai L, Cohen RE, Rubner MF (2006) Nanoporosity driven superhydrophilicity: a means to create multifunctional antifogging coatings. Langmuir 22:2856–2862
38. Dong H, Ye P, Zhong M, Pietrasik J, Drumright R, Matyjaszewski K (2010) Superhydrophilic surfaces via polymer-SiO$_2$ nanocomposites. Langmuir 26:15567–15573
39. Barthlott W, Neinhuis C (1997) The purity of sacred lotus or escape from contamination in biological surfaces. Planta 202:1–8
40. Cheng YT, Rodak DE, Wong CA, Hayden CA (2006) Effects of micro and nano structures on the self-cleaning behaviour of lotus leaves. Nanotechnology 17:1359–1362
41. Ensikat EJ, Ditsche-Kuru P, Neinhuis C, Barthlott W (2011) Superhydrophobicity in perfection: the outstanding properties of the lotus leaf. Beilstein J Nanotechnol 2:152–161
42. Li W, Amirfazli A (2008) Hierarchical structures for natural superhydrophobic surfaces. Soft Matter 4:462–466
43. Liu HH, Zhang HY, Li W (2011) Thermodynamic analysis on wetting behavior of hierarchical structured superhydrophobic surfaces. Langmuir 27:6260–6267
44. Su Y, Ji B, Zhang K, Hao H, Huang Y, Hwang K (2010) Nano to micro structural hierarchy is crucial for stable superhydrophobic and water-repellent surfaces. Langmuir 26:4984–4989
45. Yu Y, Zhao ZH, Zheng QS (2007) Mechanical and superhydrophobic stabilities of two scale surfacial structure of lotus leaves. Langmuir 23:8312–8316
46. Extrand CW (2011) Repellency of the lotus leaf: resistance of water intrusion under hydrostatic pressure. Langmuir 27:6920–6925
47. Bittoun E, Marmur A (2012) The role of multi-scale roughness in the lotus effect: is it essential for superhydrophobicity. Langmuir 28:13933–13942
48. Nosonovsky M (2007) Multiscale roughness and stability of superhydrophobic biomimetic interface. Langmuir 23:3157–3161
49. Gao L, McCarthy TJ (2006) The lotus effect explained: two reasons why two length scale of topography are important. Langmuir 22:2966–2967
50. Luo C, Zheng H, Wang L, Fang H, Hu J, Fan C, Cao Y, Wang J (2010) Direct three-dimensional imaging of the buried interfaces between water and superhydrophobic surfaces. Angew Chem Int Ed 49:9145–9148
51. Cao L, Hu HH, Gao D (2007) Design and fabrication of micro-textures for inducing a superhydrophobic behavior on hydrophilic materials. Langmuir 23:4310–4314
52. Cheng YT, Rodak DE (2005) Is the lotus leaf superhydrophobic? Appl Phys Lett 85:144101

53. Boreyko JB, Chen CH (2009) Restoring superhydrophobicity of lotus leaves with vibration-induced dewetting. Phys Rev Lett 103:174502
54. Gao L, McCarthy TJ, Zhang X (2009) Wetting and superhydrophobicity. Langmuir 25:14100–14104
55. Law KY (2012) Nanostructured coatings, surfaces and film Surf Innovations 1:75–76
56. Law KY (2014) Definitions for hydrophilicity, hydrophobicity, and superhydrophobicity: getting the basics right. J Phys Chem Lett 5:686–688
57. Liu K, Jiang L (2012) Bio-inspired self-cleaning surfaces. Annu Rev Mater Res 42:231–263
58. Yan YY, Gao N, Barthlott W (2012) Mimicking nature superhydrophobic surfaces and grasping the wetting process: a review on recent progress in preparing superhydrophobic surface. Adv Colloid Interface Sci 169:80–105
59. Sas I, Gorga RE, Joines JA, Thoney KA (2012) Literature review on superhydrophobic surfaces produced by electrospinning. J Polym Sci B Polym Phys 50:824–845
60. Nishimoto S, Bhushan B (2013) Bioinspired self-cleaning surfaces with superhydrophobicity, superoleophobicity, and superhydrophilicity. RSC Advances 3:671–690
61. Celia E, Darmanin T, de Givenchy ET, Amigoni S, Guttard F (2013) Recent advances in designing superhydrophobic surfaces. J Colloid Interface Sci 402:1–18
62. Drelich J, Marmur A (2013) Physics and applications of superhydrophobic and superhydrophilic surfaces and coatings. Surf Innovations 2:211–227
63. Oner D, McCarthy TJ (2000) Ultrahydrophobic surfaces. Effect of topography length scale on wettability. Langmuir 16:7777–7782
64. Furstner R, Barthlott W, Neinhuis C, Walzel P (2005) Wetting and self-cleaning properties of artificial superhydrophobic surfaces. Langmuir 21:956–961
65. Martines E, Seunarine K, Morgan H, Gadegaard N, Wilkinson CDW, Riehle MO (2005) Superhydrophobicity and superhydrophilicity of regular nanopatterns. Nano Lett 5:2097–2103
66. Byun D, Hong J, Saputra KJH, Lee YJ, Park HC, Byun BK, Lukes JR (2009) Wetting characteristics of insect wing surfaces. J Bionic Eng 6:63–70
67. Teare DOH, Spanos CG, Ridley P, Kinmond EJ, Roucoules V, Badyal JPS, Brewes SA, Coulson S, Willis C (2002) Pulsed plasma deposition of superhydrophobic nanospheres. Chem Mater 14:4566–4571
68. Darmainbin T, Guittard F (2011) Superhydrophobic fiber mats by electrodeposition of fluorinated poly(3,4-ethyleneoxythiathiophene). J Am Chem Soc 133:15627–15634
69. Greiner A, Wendorff JH (2007) Electrospinning: a fascinating method for the preparation of ultrathin fibers. Angew Chem Int Ed 46:5670–5703
70. Ming W, Wu D, van Benthem R, de With G (2005) Superhydrophobic films from raspberry-like particles. Nano Lett 5:2298–2301
71. Ma M, Gupta M, Li Z, Zhai L, Gleason KK, Cohen RE, Rubner MF, Rutledge GC (2007) Decorated electrospun fibers exhibiting superhydrophobicity. Adv Mater 19:225–259
72. Steele A, Bayer I, Loth E (2009) Inherently superoleophobic nanocomposite coatings by spray atomization. Nano Lett 9:501–505
73. Wang J, Liu F, Chen H, Chen D (2009) Superhydrophobic behavior achieved from hydrophilic surfaces. Appl Phys Lett 95:084104
74. Feng L, Song Y, Zhai J, Liu B, Xu J, Jiang L, Zhu D (2003) Creation of a superhydrophobic surface from an amphiphilic polymer. Angew Chem Int Ed 42:800–802
75. He B, Patankar NA, Lee J (2003) Multiple equilibrium droplets shapes as design criteria for rough hydrophobic surfaces. Langmuir 19:4999–5003
76. Bico J, Marzolin C, Quere D (1999) Pearl drops. Europhys Lett 47:220–226
77. Lafuma A, Quere D (2003) Superhydrophobic states. Nat Mater 2:458–460
78. Bormashenko E, Pogreb R, Whyman G, Bormashemko Y, Erlich M (2007) Vibration-induced Cassie-Wenzel transition on rough surfaces. Appl Phys Lett 90:201917
79. Bormashenko E, Pogreb R, Whyman G, Erlich M (2007) Resonance Cassie-Wenzel wetting transition for horizontally vibrated drops deposited on a rough surface. Langmuir 23:12217–12221
80. Liu G, Fu L, Rode AV, Craig VSJ (2011) Water droplet motion control on superhydrophobic surfaces: exploiting the Wenzel-to-Cassie transition. Langmuir 27:2595–2600

81. Xue Y, Chu S, Lv P, Duan H (2012) Importance of hierarchical structures in wetting stability on submersed superhydrophobic surfaces. Langmuir 28:9440–9450
82. Whyman G, Bormashenko E (2011) How to make the Cassie wetting state stable? Langmuir 27:8171–8176
83. Lee C, Kim CJ (2009) Maximizing the giant liquid slip on superhydrophobic microstructures by nanostructuring their sidewalls. Langmuir 25:12812–12818
84. Liu T, Sun W, Sun X, Ai H (2010) Thermodynamic analysis of the effect of hierarchical architecture of a superhydrophobic surface on a condensed drop state. Langmuir 26:14835–14841
85. Chen CH, Cai Q, Tsai C, Chen CC (2007) Dropwise condensation on superhydrophobic surfaces with two-tier roughness. Appl Phys Lett 90:173108
86. Tuteja A, Choi W, McKinley GH, Cohen RE, Rubner MF (2008) Design parameters for superhydrophobicity and superoleophobicity. MRS Bull 33:752–758
87. Tuteja A, Choi W, Mabry JM, McKinley GH, Cohen RE (2008) Omniphobic surfaces. Proc Natl Acad Sci U S A 105:18200–18205
88. Zhao H, Law KY, Sambhy V (2011) Fabrication, surface properties, and origin of superoleophobicity for a model textured surface. Langmuir 27:5927–5935
89. Tuteja A, Choi W, Ma M, Mabry JM, Mazzella SA, Rutledge GC, McKinley GH, Cohen RE (2007) Designing superoleophobic surfaces. Science 318:1618–1622
90. Choi W, Tuteja A, Chhatre S, Mabry JM, Cohen RE, McKinley GH (2009) Fabrics with tunable oleophobicity. Adv Mater 21:2190–2195
91. Fujii T, Aoki Y, Habazaki H (2011) Fabrication of super-oil-repellent dual pillar surfaces with optimized pillar intervals. Langmuir 27:11752–11756
92. Cao L, Price TP, Weiss M, Gao D (2008) Super water- and oil-repellent surfaces on intrinsically hydrophilic and oleophilic porous silicon films. Langmuir 24:1640–1643
93. Ahuja A, Taylor JD, Lifton V, Sidorenko AA, Salamon TR, Lobaton EJ, Kolodner P, Krupenkin TN (2008) Nanonails. A simple geometrical approach to electrically tunable superlyophobic surfaces. Langmuir 24:9–14
94. Kumar RTR, Mogensen KB, Boggild P (2010) Simple approach to superamphiphobic overhang nanostructures. J Phys Chem C 114:2936–2940
95. Zhao H, Park CK, Law KY (2012) Effect of surface texturing on superoleophobicity, contact angle hysteresis, and "robustness". Langmuir 28:14925–14934
96. Zheng Y, Gao X, Jiang L (2007) Directional adhesion of superhydrophobic butterfly wing. Soft Matter 3:178–182
97. Wu D, Wang JN, Wu SZ, Chen QD, Zhao S, Zhang H, Sun HB, Jiang L (2011) Three-level biomimetic rice-leaf surfaces with controllable anisotropic sliding. Adv Funct Mater 21:2927–2932
98. Yoshimitsue Z, Nakajima A, Watanabe T, Hashimoto K (2002) Effects of surface structure on the hydrophobicity and sliding behavior of water droplets. Langmuir 18:5818–5822
99. Xia D, Brueck SRJ (2008) Strongly anisotropic wetting on one-dimensional nanopatterned surfaces. Nano Lett 8:2819–2824
100. Xia D, He X, Jiang YB, Lopez GP, Brueck SRJ (2010) Tailoring anisotropic wetting properties on submicrometer-scale periodic grooved surfaces. Langmuir 26:2700–2706
101. Yang J, Rose FRA, Gadegaard N, Alexander MR (2009) Effect of sessile drop volume on wetting anisotropy observed on grooved surfaces. Langmuir 25:2567–2571
102. Zhao H, Law KY (2012) Directional self-cleaning superoleophobic surfaces. Langmuir 28:11812–11818
103. Chu KH, Xiao R, Wang EN (2010) Uni-directional liquid spreading on asymmetric nanostructured surfaces. Nat Mater 9:413–417
104. Law KY, Zhao H (2015) Design principles for robust superoleophobicity and superhydrophobicity. In Ras R, Marmur A (eds) Non-wettable surfaces, Royal Society of Chemistry, to be published.

Chapter 5
What Do Contact Angles Measure?

Abstract Contact angle measurement has widely been used to characterize the properties of solid surfaces and study liquid–surface interactions. It has been known for some time that, while the measurement itself is deceptively simple, the interpretation is not straightforward and can be very complex. Correlations between contact angle data (static contact angle θ, advancing and receding contact angle θ_A and θ_R, hysteresis $(\theta_A-\theta_R)$, and sliding angle α) and surface wettability and adhesion are at times confusing. In an attempt to find out what surface properties contact angles are measuring, Samuel, Zhao, and Law systematically measure the wetting and adhesion forces between water and 20 surfaces and correlate them with contact angle data. The surface properties of the 20 surfaces vary from hydrophilic to hydrophobic to superhydrophobic, and their morphology varies from atomic smooth to homogeneously rough in the nano- and micron scale. Based on the good correlations found between θ_A and the wetting force and θ_R with the adhesion force, it was concluded that θ_A and θ_R are measures of surface wettability and adhesion, respectively. Since sliding angle α is a measure of drop mobility, it is recommended that surface should be characterized by their wettability, adhesion and stickiness using θ_A, θ_R, and α, respectively. As for contact angle hysteresis, the analysis suggests that it is a measure of the difference in liquid–surface interfacial tension during advancing and during receding. The use of the basic concepts described in this chapter to comprehend properties displayed by some recently reported unconventional surfaces is discussed. These unconventional surfaces are surfaces with very large contact angles but very sticky or with small contact angles and very slippery.

Keywords Contact angle measurement • Static contact angle • Advancing contact angle • Receding contact angle • Sliding angle • Contact angle hysteresis • Liquid–solid interactions • Data interpretation • Wetting interaction • Adhesion interaction • Young–Dupre equation • Contact angle hysteresis mechanism

5.1 Background

Since the report of the angle of contact between a liquid droplet and a solid surface by Thomas Young more than two centuries ago [1], there has been continuous arguments in the literature regarding the validity of the Young's angle and how it may be used to study surface and liquid–surface interactions. The Young's angle by itself is

© Springer International Publishing Switzerland 2016

K.-Y. Law, H. Zhao, *Surface Wetting*, DOI 10.1007/978-3-319-25214-8_5

problematic because it consists of four quantities and two of which cannot be measured reliably. The observation of the advancing and receding contact angle in addition to the Young's angle by Rayleigh [2] and later confirmed by Bartell and coworkers [3, 4] fired up the conversation further. Researchers in the nineteenth century were strongly influenced by the thermodynamic approaches put forwarded by Dupre [5] and Gibbs [6]. As pointed out by Shuttleworth and Bailey [7] in 1948, thermodynamic (free energy) is only part of the contributor for the Young's angle during liquid wetting. This point was recognized [8] but not well understood in the literature. In view of the observation of three contact angles (static contact angle θ and advancing and receding contact angle θ_A and θ_R), some researchers created the term "ideal" surface for surfaces that are smooth, rigid, and has zero hysteresis, $\theta_A = \theta_R$. The rest, including rough and smooth heterogeneous surfaces, would be the "real" surfaces. Some believed that hysteresis is a result of defect or heterogeneity or roughness [9]. Good [10] advocated strongly that authors should either report both θ_A and θ_R or a static contact angle (θ) along with the hysteresis value ($\theta_A - \theta_R$), otherwise journals should not published these papers. The arguments seem endless and remain unsettled to date. Given this backdrop, misperceptions, such as (a) non-stickiness is a result of a large contact angle, (b) surface adhesion is due to contact angle hysteresis, or (c) contact angle θ is related to both wettability and adhesion, have been reported [11–14]. Our primary intention here is to make readers aware of the key debates, while sparing the details of all past arguments. We feel that a detailed account of all the arguments would create more harm and confusion than good. Rather, we are looking forward and articulate the basic surface concepts that are needed to advance the science for surfaces in the future.

Now let's take a step back and ask ourselves a simple question. Why are we interested in contact angle measurement? The short answer is that contact angle measurement is a very simple measurement tool for surface science. The measured contact angle is known to provide insight about surface properties as well as how liquid and surface interact, such as wetting, spreading, adhesion, and evaporation. Unfortunately, due to the difficulty in interpreting contact angle data, the literature often consists of controversial and sometimes conflicting messages. For example, the use of water as a probing liquid is very common, and surfaces with large water contact angles are highly hydrophobic, and they are usually lower in surface energy [15–17]. What's not well recognized is that large water contact angle and low surface energy may not be correlating to both wettability and adhesion simultaneously [18–21]. Tsai, Chou, and Penn [19] reported the lack of correlation between contact angle data and the adhesive performance for the smooth surfaces on a series of Kevlar fiber. Murase and Fujibayashi [20] found that while the surface of their fluorinated polymer exhibit a larger water contact angle ($117°$) than that of the polydimethysiloxane sample ($102°$), the interactive energy with water for the fluoropolymer is three times higher (~50.89 versus 15.64 mJ/m^2). Silicones and fluoropolymers, e.g., Teflon (PTFE), are two classes of well-known low-surface-energy materials that are frequently used in the fusing components in the printing industry [22, 23]. Both materials are highly hydrophobic with PTFE having a slightly larger water contact angle than PDMS silicone (112–$117°$ for PTFE as compared to 102–$103°$ for PDMS). However, PTFE was consistently shown to adhere stronger to water and has a larger sliding angle [20, 21]. To an extreme, Gao and McCarthy [24] even called Teflon hydrophilic. The fundamental question we have to ask is: what does contact angle mean to surface attraction and adhesion? Are these interactive forces in any way correlating to the surface properties? These questions are not new. They had been asked before and remained to be clarified [25]. With the advance of modern instrumentation, we thought if liquid–surface interactions can be measured directly and independently, a study of how liquid–surface interactions are correlating to the different contact angles would be useful and interesting. It is our hope that examination of these correlations or trends would shine light on the true meaning of each contact angle measurement.

5.2 Contact Angles and Liquid–Solid Interactions

5.2.1 Contact Angles

The most common properties for surfaces are their wettability and adhesion. Wettability is used to describe the interaction when a liquid first makes contact to a solid surface. On the other hand, adhesion is a description of the force when a liquid is separating from a solid surface. These are two very different interactions, yet contact angles have been used to study and correlate both [25]. Therefore, it comes no surprise that controversial and conflicting findings exist. To find out what contact angles are measuring, Samuel, Zhao, and Law [26] recently launched a systematic study involving measurements of all the contact angles for a series of surfaces and correlating them with measurable forces derived from wetting and adhesion. Water was chosen as the probing liquid. The static and dynamic contact angles (θ, θ_A, θ_R, and sliding angle α) and contact angle hysteresis on 20 different surfaces (**1–20**) were determined. The data are summarized in Table 5.1. These 20 surfaces were cleaned appropriately prior to all contact angle measurements. They represent surfaces of all traits. Their affinity toward water varies from hydrophilic to hydrophobic to superhydrophobic. Their surface morphology varies from atomic smooth (self-assembled monolayers or CVD films on silicon wafer, **7–11**) to films from blade coating (**1–3, 6, 13**) to commercial plastic films (**4, 5, 12, 14**) to photolithographic textured surfaces (**15–18**) to rough surfaces from nature (**19, 20**). Figures 5.1, 5.2, and 5.3 plot the static contact angle θ against sliding angle α, contact angle hysteresis (θ_A–θ_R), and ($\cos\theta_R$–$\cos\theta_A$), respectively.

In the literature, sliding angle and contact angle hysteresis have been thought to be related to surface stickiness and adhesion. The larger the sliding angle, the stickier the surface; the larger the contact angle hysteresis, the stronger the surface adhesion. Now, if one considers all 20 surfaces, flat, smooth, rough, and textured, the data points are very scattered. The results lead to the general conclusion that, there is no correlation between static contact angle θ and surface stickiness or adhesion. Although similar conclusions have been reached earlier [15, 16, 19, 20], the lack of correlation observed by Samuel, Zhao, and Law [26] is still significant because the study comprised of a very wide range of surfaces, implying that the noncorrelation is genuine.

Table 5.1 Contact angle measurement data and wetting and adhesion force data for water on different surfaces (data from [26])

No	Polymer surfaces	θ[a]	θ_A[b]	θ_R[c]	α[d]	$(\theta_A - \theta_R)$[e]	Snap-in force (µN)	Pull-off force (µN)
1	Polyurethane (PU)	70.5°	85°	48.9°	51°	36.1°	471.4±14.2	179.4±2.2
2	PU—2 % Silclean	98.2°	104.3°	76.3°	31°	28°	316.9±17.7	175.7±3.2
3	PU—8 % Silclean	104.3°	105.9°	88.1°	23°	17.8°	292.3±7.3	172.6±2.0
4	Polyimide	80.1°	82.5°	56.1°	26.4°	26.4°	442±54.4	167.3±10.3
5	Plexiglass	86.5°	93.9°	77.3°	29.1°	16.6°	387.6±13.4	157.4±8.1
6	Polycarbonate	92.4°	98.2°	68.1°	59.2°	30.1°	338.7±22.4	163.2±1.9
7	i-CVD silicone	87.9°	91.2°	62.2°	f	29°	379±14.2	175.5±2.6
8	i-CVD fluorosilicone	115.9°	118°	90.3°	18.2°	27.7°	141±6.3	148.7±1.1
9	i-CVD PTFE	127.7°	134.9°	73.6°	f	61.3°	72.4±4.2	168.8±1.5
10	OTS SAM	109°	117.4°	94.6°	13°	22.8°	197±9.5	141.3±0.2
11	FOTS SAM	107.3°	116°	95°	13.6°	21°	226.6±8.7	139.5±1.8
12	Perfluoroacrylate	113°	113.1°	61°	f	52.1°	398.7±18.7	149.2±12.6
13	Hydrophobic sol gel	112.2°	111.6°	92.4°	5.6°	19.2°	197.9±12.2	111.4±8.3
14	PTFE	117.7°	126.6°	91.9°	64.3°	34.7°	89.7±15.4	162.0±6.6
Textured silicon surface (pillar diameter/spacing)								
15	3/4.5 µm	149°	160°	130.8°	20°	29.2°	0	29.4±2.6
16	3/6 µm	156.2°	161.3°	142.6°	10.1°	18.7°	0	15.6±1.7
17	3/9 µm	154.1°	159.9°	148.9°	5.7°	11°	0	8.54±0.8
18	3/12 µm	156.2°	160.8°	151.8°	3.4°	9°	0	4.71±0.7
19	Rose petal—front	144.7°	150.7°	131.6°	6.1°	19.1°	2.9±1.3	31.5±8.0
20	Rose petal—back	132.4°	136.6°	85.7°	f	50.9°	23.4±5.7	140.0±15.8

[a]Static contact angle, estimated error <2°
[b]Advancing contact angle, estimated error <2°
[c]Receding contact angle, estimated error <2°
[d]Sliding angle, estimated error <3°
[e]Contact angle hysteresis
[f]Water droplets do not slide and are struck at 90° tilt angle

On the other hand, if one just focuses on surface **15–20** in Figs. 5.1, 5.2, and 5.3 (labeled as *open diamond*), there may exist reasonable correlations between the static contact angle and sliding angle and hysteresis. These are all superhydrophobic surfaces. The good correlations may merely indicate that if a narrow range of material property is considered, a fortuitous good correction may be obtained.

5.2.2　Wetting and Adhesion Force Measurements

To gain a better understanding of what contact angles are measuring, Samuel, Zhao, and Law [26] measured the wetting (attractive) and adhesion force between water and surface **1–20** directly using a microbalance inside a tensiometer. A schematic of the measurement procedure is shown in Fig. 5.4, and details of the instrumentation have given elsewhere [26].

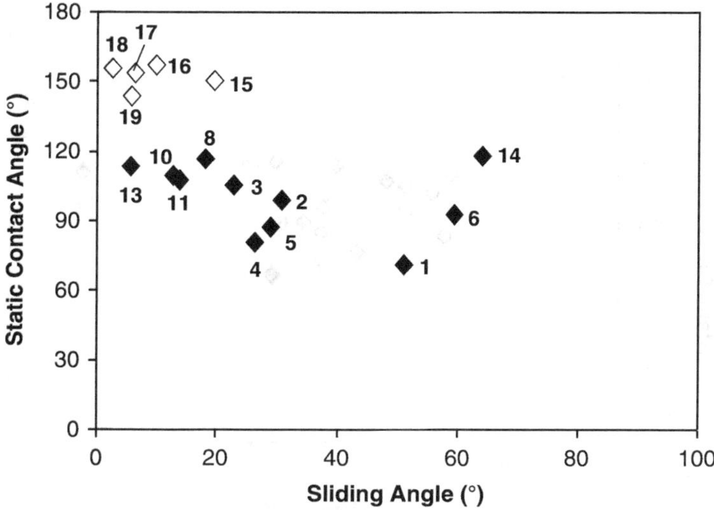

Fig. 5.1 Plots of static contact angle versus sliding angle for flat surfaces **1–14** and rough surfaces **15–19** (reproduced with permission from [26], Copyright 2011 The American Chemical Society)

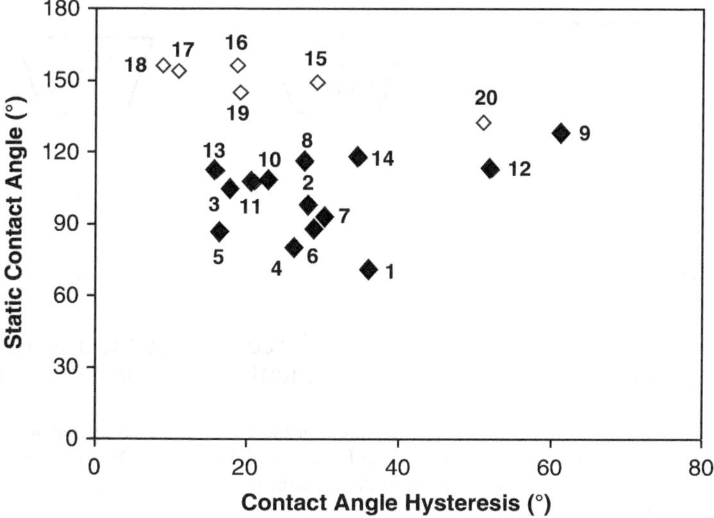

Fig. 5.2 Plots of static contact angle versus contact angle hysteresis ($\theta_A - \theta_R$) for flat surfaces **1–14** and rough surfaces **15–20** (reproduced with permission from [26], copyright 2011 The American Chemical Society)

Briefly, a 5 mg water droplet was first deposited onto a platinum ring, which is attached to a microbalance. The surface of interest is placed on a stage where it can move up and down steadily at a slow rate (10 μm/s) by a computer-controlled stepping motor. Prior to the measurement, the microbalance is set to zero. When the water drop first "touches" the surface (Fig. 5.4 step b), an attractive snap-in force

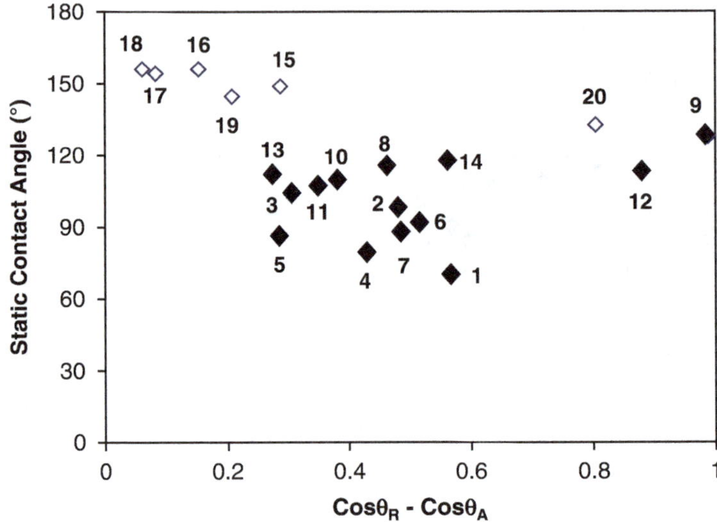

Fig. 5.3 Plots of static contact angle versus $\cos\theta_R$–$\cos\theta_A$ for flat surfaces **1–14** and rough surfaces **15–20** (Reproduced with permission from [26], copyright 2011 The American Chemical Society)

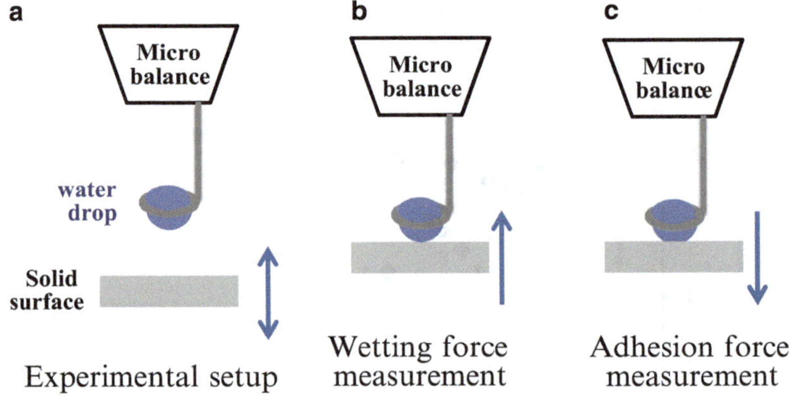

Fig. 5.4 Schematic of the apparatus and procedures for measuring the wetting and adhesion inter-actions between water and various surfaces. (**a**) Experimental setup. (**b**) Wetting force measure-ment. (**c**) Adhesion force measurement (reproduced with permission from [62], copyright 2014 The American Chemical Society)

attributable to the wetting interaction between the water droplet and the surface is recorded. The stronger the water–surface attraction, the stronger the snap-in force. After the water droplet and the surface made contact, the stage is retracted slowly; a pull-off force is recorded when the water droplet and the surface separate (Fig. 5.4 step c). It is worthy pointing out that the measurement of the pull-off force is a little bit more complicated. When the separation between the water droplet and the surface is clean where no residual water droplet is observed after the separation, the pull-off force should correspond to the adhesion force between the water droplet and the surface at vertical separation. However, when a residual water droplet is observed on the surface after the pull-off, it implies that the adhesion between the surface and the water droplet is stronger than the cohesive force within the water droplet. What's recorded is the force when the water droplet breaks. The magnitude of this force is related to the adhesion between the water droplet and the surface, cohesive force of the water droplet (surface tension), and the contact area when the water droplet breaks. This is actually a very important observation. The implication of the resid-ual droplet to surface definition will be discussed further in the next chapter in this book. Here, we assume that the measured force is still dominated by the adhesion force between the water droplet and the surface. Both snap-in force and pull-off force data for the water droplet on surfaces **1–20** are included in Table 5.1.

5.2.3   Wetting Interaction and Contact Angles

Numerous attempts were made to correlate the contact angle data from surfaces **1–20** with the force data. The snap-in force, which measures how strongly the water droplet and the surface interact, is shown to correlate well to θ_A. Figure 5.5 plots the snap-in force on surfaces **1–20** versus θ_A. The snap-in force increases monotonously

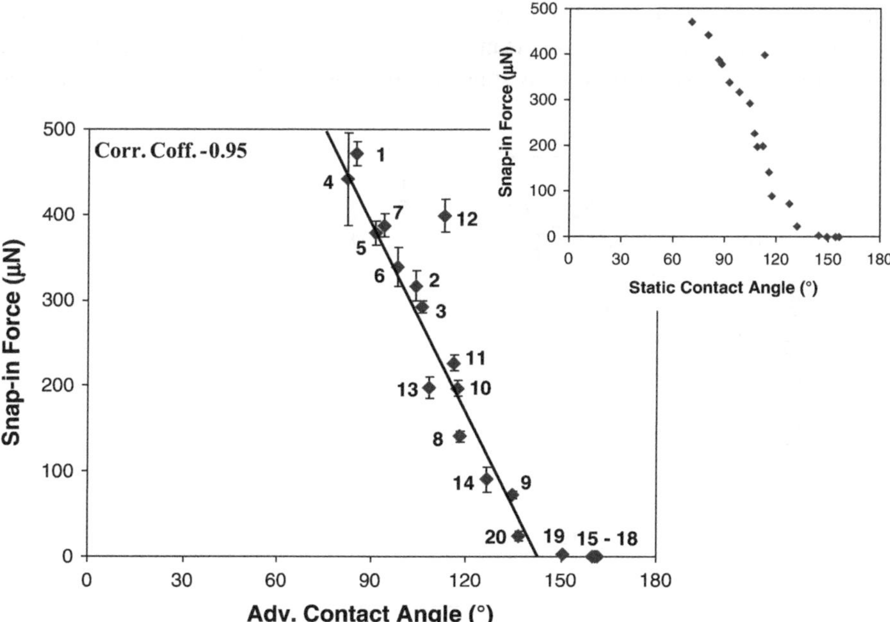

Fig. 5.5 Plot of snap-in force versus advancing contact angle (*inset*: plot of snap-in force versus static contact angle) (reproduced with permission from [26], copyright 2011 The American Chemical Society)

as θ_A decreases. Since snap-in force measures the strength of the water attraction, the correlation indicates that θ_A correlates to surface wettability; the larger the θ_A value, the lower the surface wettability. The snap-in force becomes negligible at $\theta_A \geq 145°$, indicating that there is practically no attractive interaction between water and surfaces 15–19. The significance of this observation will be discussed in Chap. 6. Incidentally, a reasonably good correlation is also obtained when the snap-in force is plotted against θ (inset of Fig. 5.5). The observation is not unexpected since static contact angle θ always tracks well with the advancing contact angle θ_A due to the way the two measurements are made.

5.2.4 Adhesion Interaction and Contact Angles

Sliding angle α is the tilt angle when the sessile droplet starts to move on an inclined surface. It measures stickiness of the surface or mobility of the liquid droplet; the stickier the surface, the larger the sliding angle and the lower the drop mobility. As a result, a number of reports suggested that α, contact angle hysteresis $(\theta_A–\theta_R)$, and $(\cos\theta_R–\cos\theta_A)$ are related to surface adhesion [11, 27–30]. Since the pull-off force measures the adhesion between the water droplet and the surface, we can then test these hypotheses by plotting the pull-off force against α, $(\theta_A–\theta_R)$, and $(\cos\theta_R–\cos\theta_A)$. The results are shown in Figs. 5.6, 5.7, and 5.8, respectively. In these plots, the surfaces are further classified into two groups. Specifically, surfaces labeled with *open squares* are surfaces that are clean after pull-off (no residual water droplet), and *solid squares* are surfaces with a small residual water droplet left behind after

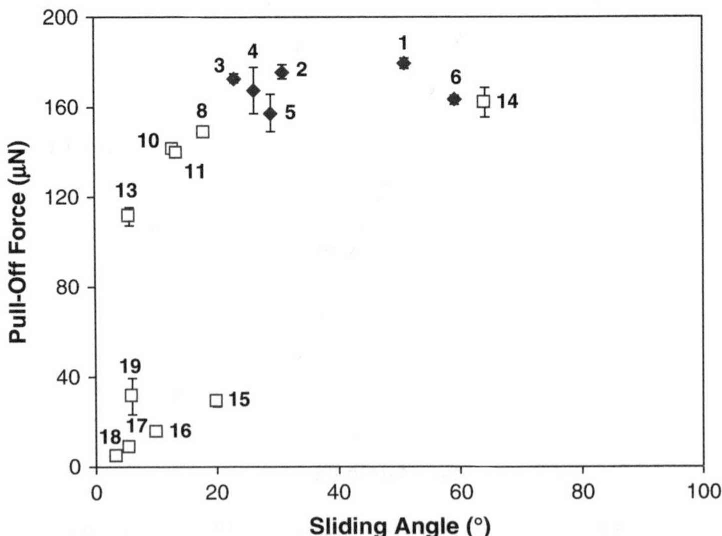

Fig. 5.6 Plot of the pull-off force for surfaces **1–20** versus sliding angle α (reproduced with permission from [26], copyright 2011 The American Chemical Society)

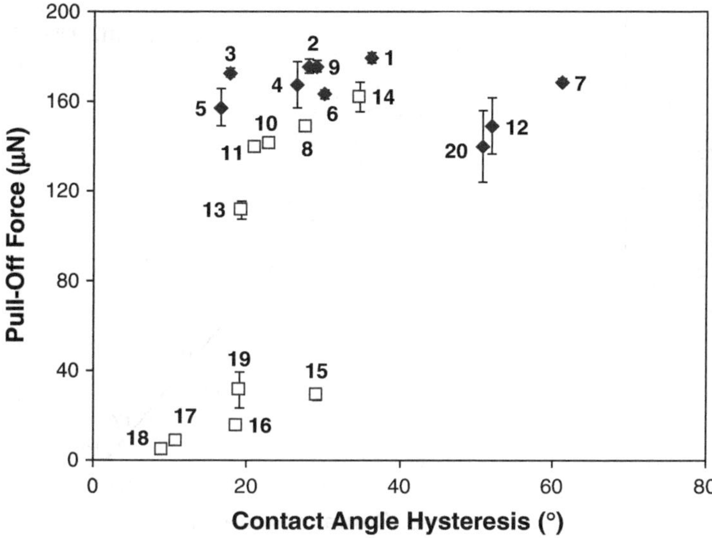

Fig. 5.7 Plot of the pull-off force for surfaces **1–20** versus contact angle hysteresis (θ_A–θ_R) (reproduced with permission from [26], copyright 2011 The American Chemical Society)

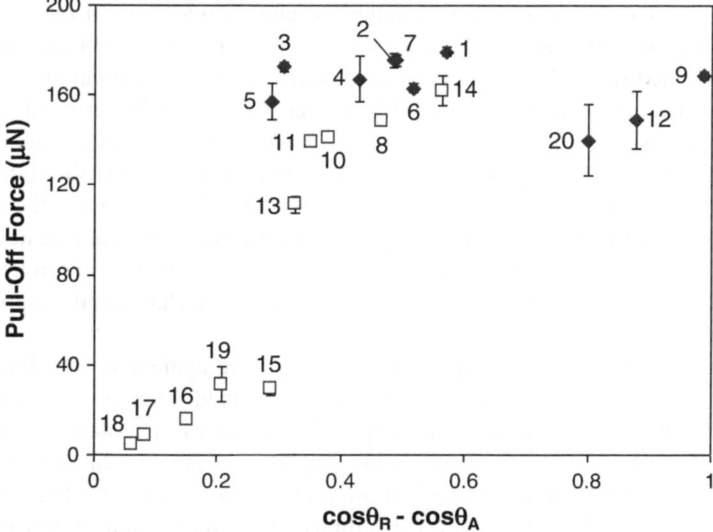

Fig. 5.8 Plot of the pull-off force for surfaces **1–20** versus $\cos\theta_R$–$\cos\theta_A$. (Reproduced with permission from [26], copyright 2011 The American Chemical Society)

pull-off. Examination of the data suggests that there is little correlation between the pull-off force on surfaces **1–20** with the sliding angle α, contact angle hysteresis (θ_A–θ_R), and ($\cos\theta_R$–$\cos\theta_A$). The absence of any correlation implies that α, (θ_A–θ_R), and ($\cos\theta_R$–$\cos\theta_A$) are not measures of surface adhesion. On the other hand, further

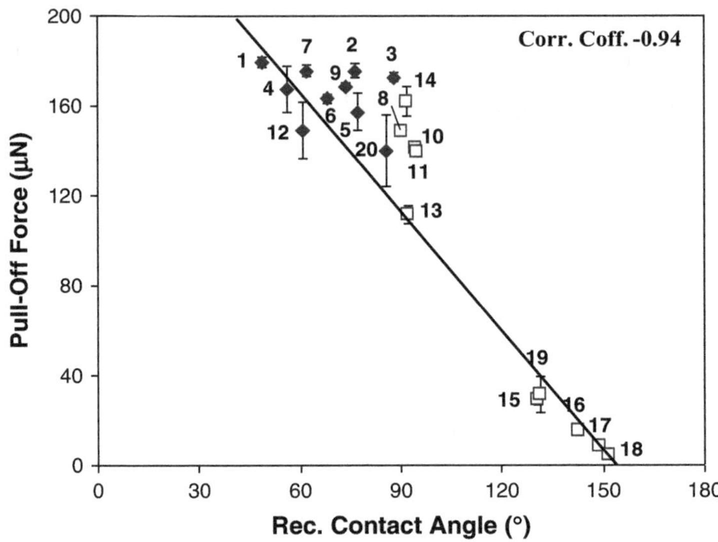

Fig. 5.9 Plot of the pull-off force versus receding contact angle from surfaces **1–20** (reproduced with permission from [26], copyright 2011 The American Chemical Society)

examination of the data reveals that some correlation may exist if one groups the surfaces into two different sets. The first set of surfaces includes **1–14** and **20**. Similar to the combined set of data, this group of surfaces shows no correlation with sliding angle α, contact angle hysteresis (θ_A–θ_R), and ($\cos\theta_R$–$\cos\theta_A$). The second set of surfaces, **15–19**, are all superhydrophobic. The water droplet is in the Cassie–Baxter state on these surfaces. The pull-off force seems to correlate well to the sliding angle α, contact angle hysteresis (θ_A–θ_R), and ($\cos\theta_R$–$\cos\theta_A$). The characteristic of this set of surfaces is that water pulls off cleanly from these surfaces and they all have a relatively low pull-off force. Of course, one can always argue that this correlation only occurs over a narrow range of contact angles. Or this correlation only happens with rough surfaces.

On the other hand, when the pull-off force is plotted against the receding contact angle of surfaces **1–20**, a good correlation is obtained (Fig. 5.9). Pull-off force increases monotonically as the receding angle decreases. The data appears to be more scattered at smaller receding contact angles, presumably due to the involvement of different "separating states" in the pull-off measurements. The water drop may break differently depending on specificity of the interaction. In any event, the result in Fig. 5.9 suggests that receding contact angle correlates well to surface adhesion. Another significant observation in Fig. 5.9 is that there appears to be a cutoff at receding angle 90° where surfaces are clean and without any residual water droplet after pull-off when θ_R is >90°. This point will be discussed further in Chap. 6.

The work of adhesion (W_h) between a liquid droplet and the surface can be quantified as shown in (5.1) according to the surface literature [24, 30].

$$W_h = \gamma_{LV}\left(1 + \cos\theta_R\right)$$

$$(5.1)$$

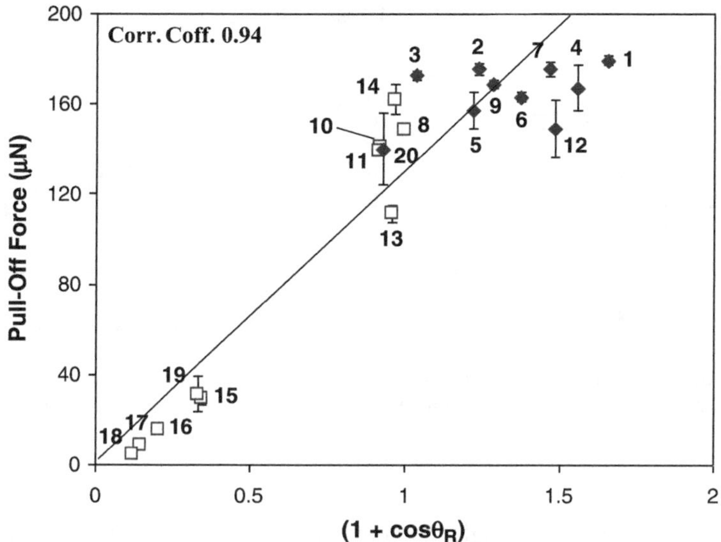

Fig. 5.10   Plot of pull-off force measured on the tensiometer versus $(1+\cos\theta_R)$ (reproduced with permission from [26], copyright 2011 The American Chemical Society)

Thus, if the pull-off force is a measure of surface adhesion, it should correlate to $(1+\cos\theta_R)$. Indeed, a good linear correlation is obtained in Fig. 5.10 when the pull force is plotted against $(1+\cos\theta_R)$. The overall result confirms that the pull-off force indeed measures the adhesion between the water droplet and the surface when they separate during the experiment in the tensiometer. More importantly, the plot in Fig. 5.10 establishes convincingly that receding angle θ_R indeed measures adhesion. The fundamental reason why θ_R is correlating to surface adhesion is because θ_R is dictated by the ability of the liquid droplet to de-pin at the liquid–solid interface; the smaller the θ_R, the more difficult it is to de-pin and the stronger the adhesion interaction.

5.3   Sliding Angle

When the sessile droplet on a horizontal surface is tilted, the normally circular droplet is distorted or elongated by gravity. The degree of distortion is related to the contact angle hysteresis and the angle of tilting [27, 28]. Figure 5.11a depicts a schematic of a distorted sessile droplet on an inclined surface and the forces that act on the liquid droplet on the tilted surface. The driving force (F) for a liquid droplet to slide is gravity and is given by

$$F = mg \cdot \sin\alpha \tag{5.2}$$

where m is the mass of the liquid droplet and g is the gravitational constant.

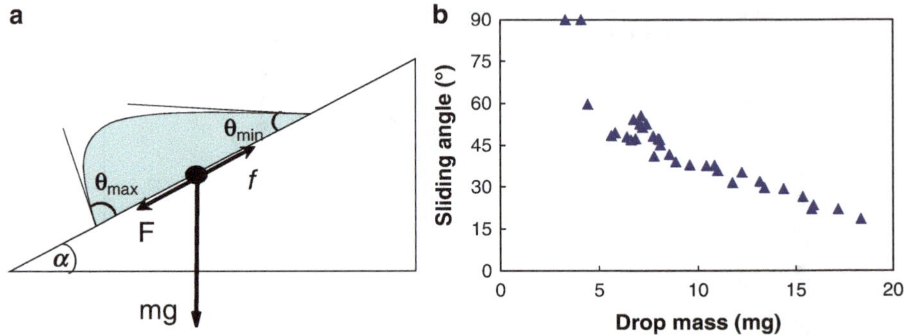

Fig. 5.11 (**a**) Schematic of a distorted sessile drop on a tilted surface and (**b**) plot of sliding angle as a function of drop mass for hot polyethylene wax droplets on FOTS-treated silicon wafer

Sliding angle α is the tilt angle moment before drop sliding. Thus, α is a measure of surface stickiness or drop mobility. For a given drop mass, surface stickiness increases as α increases. Figure 5.11b shows a typical decrease profile for the sliding angle as a function of drop mass. It is therefore important to control and know the drop mass (or volume) when comparing surface stickiness or drop mobility from sliding angle data.

The retention (or frictional) force (f) that keeps the drop from sliding is a lot more difficult to determine. The schematic in Fig. 5.11a is just a simplified view of the situation. In short, the sessile droplet is distorted three dimensionally. Figure 5.12a shows a photograph of a water drop on a tilted surface. The smallest angle of contact for the distorted drop is at the trail edge and is designed as $\theta_{\min}$. The angle of contact increases along the contact line downward; the largest angle is observed at the lead edge and is designated as $\theta_{\max}$ [31]. A 2D top view showing the change in the angle of contact around the three-phase contact line of the distorted droplet is given in Fig. 5.12b. Principally, the retention force is the sum of all forces around the circumstance of the distorted droplet [32]. To make the problem manageable, Furmidge [29] and Macdougall and Ockrent [30] simplified the complex 3D geometry to 2D, and f can be expressed as

$$f = \gamma_{\mathrm{LV}} \cdot R \cdot k \cdot \left(\cos\theta_{\min} - \cos\theta_{\max} \right) \tag{5.3}$$

where γ_{LV} is the surface tension of the liquid, R is the length scale for the contour of the drop, and k is an adjustable parameter based on experimental data.

Just before drop sliding on the tilted surface, $F=f$. Krasovitski and Marmur [33] equated the sliding angle to $\theta_{\max}$ and $\theta_{\min}$ as

$$\sin\alpha = C \cdot \gamma_{\mathrm{LV}} \cdot \left(\cos\theta_{\min} - \cos\theta_{\max} \right) \tag{5.4}$$

where C is a constant that includes the gravitational acceleration, density of the liquid, and the geometric parameters of the drop.

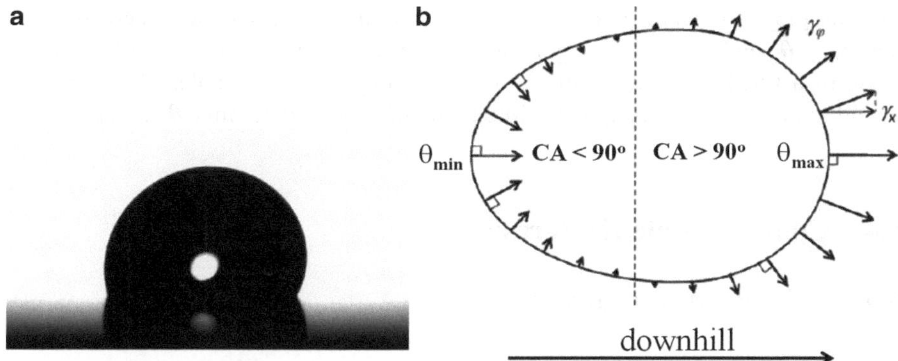

Fig. 5.12 (**a**) A photograph of a water droplet on a tilted surface and (**b**) schematic showing the forces acting on the sessile drop on the tilted surface (**b**, reproduced with permission from [32], copyright 2008 Elsevier)

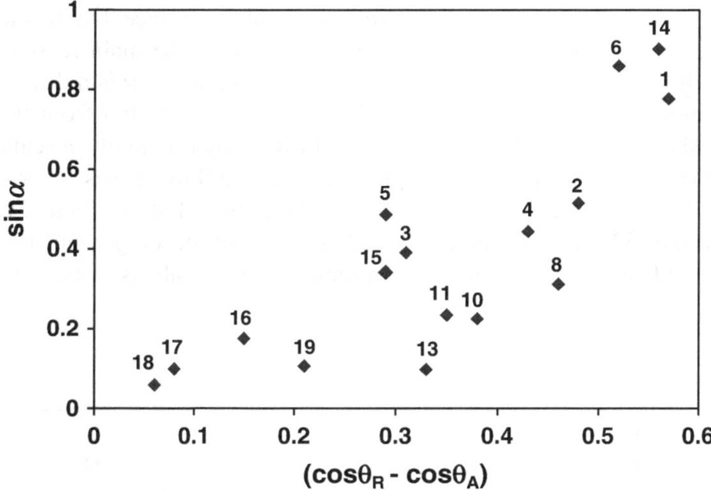

Fig. 5.13 Plot of sinα for surfaces in Table 5.1 versus $(\cos\theta_R-\cos\theta_A)$

In the surface science literature, it was shown as early as 1942 that the lead edge and trail edge angles on the tilted plane are θ_A and θ_R, respectively [30]. More recently, ElSherbini and Jacobi [34] reported that θ_{min} and θ_{max} are approximately equal to θ_R and θ_A for a large number of liquid–surface combinations. If that is true, plot of sinα versus $(\cos\theta_R-\cos\theta_A)$ (data in Table 5.1) should yield a linear relationship. Experimentally, a very scattered plot is obtained in Fig. 5.13. Although there is a very rough trend, the lack of a good linear correlation suggests that $\theta_{max} \neq \theta_A$ and $\theta_{min} \neq \theta_R$. This is not a surprising outcome because many approximations have been used to derive at Eq. (5.4). After all, both Pierce et al. [32] and Krasovitski and

Marmur [33] demonstrated that there is only a limited contact angle hysteresis range that $\theta_{max}=\theta_A$ and $\theta_{min}=\theta_R$. The general assumption that the lead edge and trial edge angle of a tilted liquid drop are θ_A and θ_R is highly questionable. Therefore, we would not recommend using the titling plane method to determine θ_A and θ_R.

5.4   Contact Angle Hysteresis

5.4.1   What Does It Measure?

Contact angle hysteresis defines as the difference between θ_A and θ_R. Results in Figs. 5.2 and 5.7 show that it does not correlate to the static contact angle θ and receding angle θ_R. This indicates that contact angle hysteresis is neither measuring surface wettability nor adhesion directly. Similar conclusion was also reached by Della Volpe et al. [35], who concluded that contact angle hysteresis is completely independent of the hydrophobicity/hydrophilicity of the surface. On the other hand, many researchers have suggested that low hysteresis is the main reason for high drop mobility on tilted surfaces [36–39]. Since sliding angle α is a direct measure for drop mobility, a plot of sliding angle α versus (θ_A and θ_R) from the data in Table 5.1 should be a good test for the hypothesis. Experimentally, a scattered plot (Fig. 5.14) is obtained. The scattered plot suggests that if hysteresis is correlating to drop mobility, the correlation is only qualitative at best. Indeed, Pierce, Carmona, and Amirfazli [32] also concluded that advancing and receding contact angles are not good predictors for drop mobility in their detailed analysis of the complicated

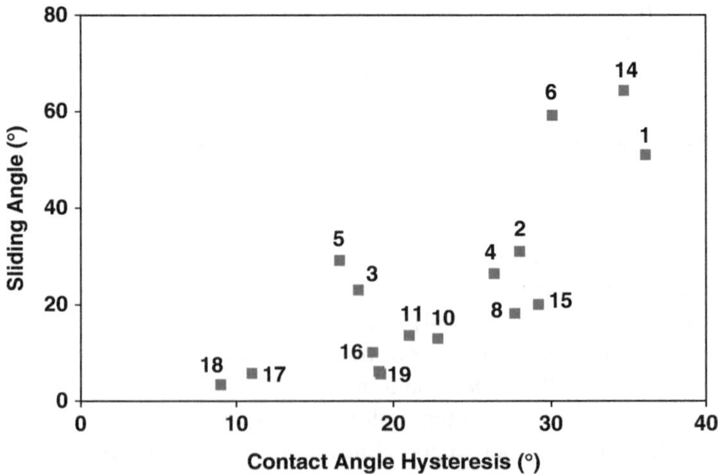

Fig. 5.14 Plot of sliding angle α for surfaces in Table 5.1 versus contact angle hysteresis

relationship among sliding angle, advancing and receding contact angle, and hysteresis.

As pointed out earlier in Chap. 3, both θ_A and θ_R are angles from their respective metastable wetting states. Their equivalent Young's equations can be written as

$$\cos\theta_A = \frac{\gamma_{SV} - \gamma_{SL}^{ad}}{\gamma_{LV}} \tag{5.5}$$

$$\cos\theta_R = \frac{\gamma_{SV} - \gamma_{SL}^{re}}{\gamma_{LV}} \tag{5.6}$$

where γ_{SV} and γ_{LV} are the surface tensions for the solid and liquid, respectively, γ_{SL}^{ad} is the liquid–solid interfacial tension during advancing, and γ_{SL}^{re} is the liquid–solid interfacial tension during receding.

During receding, the area underneath the drop is already wetted. The contact line is receding from an interface where liquid molecules and solid–surface materials are already relaxed and are in their equilibrated state. This may not be the case during advancing. Liquid molecules and materials on the solid surface may not have enough time to equilibrate at the interface as the contact line is kept advancing. In other words, there is a difference in liquid–solid interfacial tension in the advancing mode and the receding mode (γ_{SL}^{ad} versus γ_{SL}^{re}). Therefore, contact angle hysteresis is actually a measure of the difference in liquid–solid interfacial tension during advancing and during receding; the larger the difference, the larger the hysteresis.

5.4.2   Mechanisms for Contact Angle Hysteresis

Equations (5.5) and (5.6) clearly show that contact angle hysteresis originates from the difference in liquid–solid interfacial tension during advancing and during receding. The difference in wetting states during advancing and receding was mentioned by Macdougall and Ockrent [30] and by Pease [40] earlier. Historically, contact angle hysteresis was attributed to roughness and heterogeneity [7, 9, 41, 42]. Indeed, when a liquid is in the Wenzel state, it wets the rough surface fully and results in pinning of the liquid droplet on the surface and very large hysteresis [43, 44]. Extrand [45] further suggested that the geometry of the roughness may be more important than roughness alone in determining the contact angle and hysteresis of ultraphobic pillar array surfaces. It thus appears that roughness is not the only contributor to contact angle hysteresis.

The occurrence of sizable contact angle hysteresis on perfectly smooth surfaces was reported as early as 1952 by Bartell and Bjorklund [46] in a three-liquid, three-phase system comprising mercury, benzene, and water. A strong conclusion was not drawn that time presumably due to the observation of an unexplained, aging phenomenon on the three-liquid system. Four decades later, Chen and coworkers [47] reported a study of the contact angle hysteresis of several monolayer modified mica sheets with water and observed sizable hysteresis also despite working with atomic smooth samples. They attributed the large hysteresis to the relaxation of liquid molecules after the liquid wets the monolayer–mica surfaces. Rearrangement of functional groups in polymer chains and dipoles in liquid after wetting was reported at the water–hydrogel interfaces, which lead to large hysteresis [48]. Lee and coworkers demonstrated the existence of interactions between liquid molecules and dipoles on different fluoropolymer surfaces during their wettability study [49]. Extrand and Kumagai [50] designed an experiment consisting of six different surfaces (silicon wafer plus five different polymers) and five different liquids, to test whether roughness or chemical interaction at the interface are playing a role in determining the contact angle hysteresis. The 6×5 matrix includes liquid–surface combinations with a wide range of contact angle and hysteresis. They concluded that chemical interaction at the liquid–solid interface is a key contributor to the large hysteresis observed in their study. It thus becomes apparent that both chemical and physical interactions can contribute to contact angle hysteresis. Before the liquid wets the solid surface, molecules in liquid droplet and on the surface of the solid are in their respective thermodynamically stable state. As soon as the liquid wets the surface, both liquid molecules and segments of materials on the solid surface can rearrange to their relaxed state at the liquid–solid interface. The degree of relaxation will depend on the specific liquid–solid system. Surface roughness can magnify the interaction. Using the methodology of Cassie–Baxter, the total contact angle hysteresis (CAH^{tot}) for a given liquid–solid system can be defined as

$$CAH^{tot} = f_s \cdot CAH^{chem} + (1 - f_s) \cdot CAH^{rough} \tag{5.7}$$

where f_s is the solid area fraction on the rough surface, CAH^{chem} is the CAH from the chemical interaction, and CAH^{rough} is the CAH from the rough structure.

For smooth surfaces, $f_s = 1$, the chemical interaction between molecules in the liquid and the solid surface (CAH^{chem}) becomes the sole contributor. For rough surfaces, both CAH^{chem} and CAH^{rough} are contributing to the hysteresis observed. Generally speaking, CAH^{tot} for rough surface increases as f_s increases due to the pinning effect [51, 52]. For surfaces similar to the Lotus leaf, f_s approaches zero and CAH^{rough} dictates the size of the hysteresis. Factors that govern CAH^{rough} have not been well studied. Many investigators have been using measured roughness factors, such as Ra and Rz, to correlate contact angle data and usually not very fruitful. In any event, if one considers the interaction between the liquid and the rough structure at the molecular level, one would intuitively expect that the pinning geometry at the three-phase contact line should be paramountly important. Indeed, Extrand found in his study of the wetting of pillar array surfaces that the decrease of receding contact angle and increase of hysteresis is more sensitive to the geometry of the pillar than the height [45]. Very recently, Kanungo and coworkers reported that the hystereses of PDMS surfaces with bumps are larger than those with cavities for the same roughness factor in their study of the wetting of water on rough PDMS surfaces [53]. To an extreme, rough surfaces with vertical protrusion, even in the nanoscale,

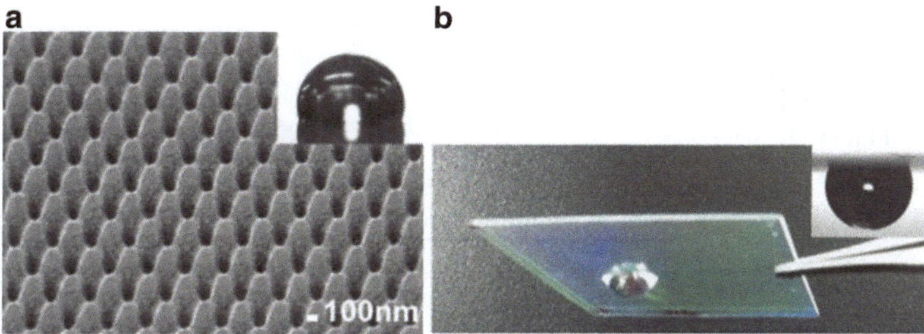

Fig. 5.15 (**a**) SEM micrograph of a polycarbonate film with nanosized protrusion (pitch and height are ~300 nm) and (**b**) polycarbonate film flipped 180° (*insets*: images of the sessile droplets) (reproduced with permission from [54], copyright 2014 The American Chemical Society)

are found to be very effective in immobilizing liquid droplets due to pinning. This of course results in a very large contact angle hysteresis. For instance, Law et al. [54] recently prepared a series of polycarbonate films with nanosized protrusion by the nanoimprinting technique and found that water droplet basically pins on the rough surface during wetting. Figure 5.15a shows a SEM micrograph of a nano-patterned polycarbonate surface (pitch and height are ~300 nm) and the water sessile droplet image. The water contact angle was at 108°. The contact angle is larger than that of a smooth polycarbonate surface (~92°), indicating that water is in the Wenzel state on the nano-patterned surface. Indeed, water droplet was found stuck and pinned when the surface is flipped 180° (Fig. 5.15b). The nano-patterned surface is thought to be useful in preventing the so-called coffee ring stain effect during inkjet printing. Similarly, Mettu, Kanungo, and Law [55] also observed analogous pinning effect on a biaxial-oriented polypropylene (BOPP) substrate. Nanosized vertical protrusions were shown to form upon heating, and the resulting protrusions pin droplets of the inkjet materials on the heated BOPP surface.

5.5 Surface Characterization Recommendations

Contact angle measurement has been a very popular tool to characterize the property of solid surfaces and understand liquid–solid interactions. Based on the data summarized in this chapter (Figs. 5.1, 5.2, 5.3, 5.4, 5.5, 5.6, 5.7, 5.8, 5.9, 5.10, 5.11, 5.12, 5.13, and 5.14), we recommend that surface should be characterized by its advancing contact angle, receding contact angle, and sliding angle. They measure surface wettability, adhesion, and stickiness or slipperiness, respectively (Fig. 5.16). In theory, static contact angle is not a measure of anything. However, researchers often found θ to correlate to the advancing angle and wettability (e.g., Fig. 5.5). This is probably due to the way both measurements are made. Specifically, both static and advancing angles involve contact line advancing prior to capturing and analyzing of

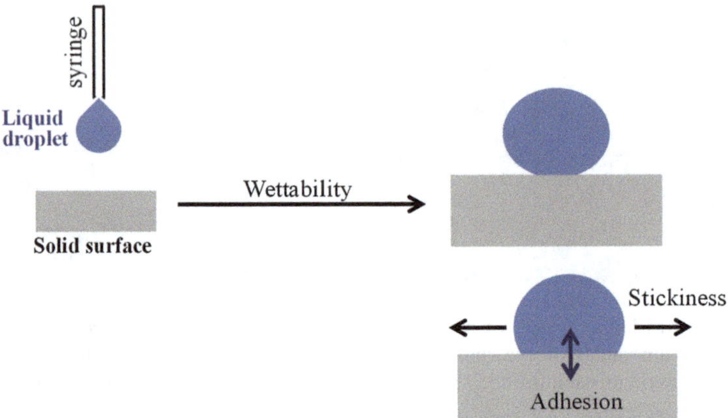

Fig. 5.16 A summary of surface characterization recommendations

the drop profile. From the liquid–solid interaction perspective, advancing contact angle will give indication if a liquid will wet or repel by a surface. Receding contact angle will allow prediction of the strength of the liquid–solid adhesion, while sliding angle will offer clue about mobility of the liquid droplet.

As discussed previously, contact angle hysteresis has been said to correlate to adhesion and drop mobility in the literature. We have to emphasize that, from the data in Figs. 5.7 and 5.14, the correlation is qualitative at best. The main reason for the existence of a rough correlation is because of the θ_R term in contact angle hysteresis. While it may be a surprise to some, hysteresis is not related to wettability either [35]. Undoubtedly more work is needed to understand the true role of contact angle hysteresis in surface characterization and wetting, Eqs. (5.5) and (5.6) clearly indicate that it originates from the difference in liquid–solid interfacial tension during advancing and during receding. In other words, the liquid–solid interface can be in two different states depending on whether it is in the advancing or receding mode. It is only with this understanding that we can comprehend some of the unexpected results reported in recent surface literature, e.g., surfaces with both small contact angle and sliding angle. These surfaces will be part of the conversation in the next section.

5.6   Myths in Adhesion and Contact Angle Hysteresis

There are many myths in surface science. One of them is the belief that large contact angle would lead to low adhesion, low hysteresis, and nonstickiness. The report of the self-cleaning effect displayed by the Lotus leaf further exacerbates that belief [56]. In any event, the self-cleaning effect did inject excitement in surface science and engineering. Studies of superhydrophobicity and more recently superoleophobicity have been fierce and intense. Numerous potential applications, e.g., self-cleaning textiles; oil- and soil-resistant clothing; anti-smudge surface for iPhone

Table 5.2 Contact angle data highlighting the expected and unexpected surface properties reported in recent literature

Surface	Liquid	θ[a]	α[b]	θ_A[c]	θ_R[d]	θ_A–θ_R[e]	References
i. Lotus leaf	Water	162°	4°				[56]
ii. FOTS-coated textured Si wafer (~3 µm/12 µm/~7 µm/wavy)[f]	Water	156°	3.4°	161°	152°	9°	[51]
	Hexadecane	154°	3.7°	162°	142°	20°	
iii. FOTS-coated textured Si wafer (~3 µm/6 µm/~7 µm/smooth)[f]	Water	152°	12°	162°	135°	27°	[57]
	Hexadecane	120°	Not slide	129°	~0°	~129°	
iv. Hydrophobized SU8 texture on Si wafer (20 µm/32 µm/30 µm)[g]	Water	–	Not slide	140°	~0°	~140°	[44]
v. OTS SAM on Si wafer[a]	Water	109°	13°	117°	95°	22°	[57]
	Hexadecane	40°	8°	45°	34°	10°	
vi. FOTS SAM on Si wafer[b]	Water	107°	13°	116°	95°	21°	[57]
	Hexadecane	73°	9°	75°	65°	10°	
vii. PDMS^{2K} on glass[h]	Water	–	–	104°	102°	2°	[58]
viii. PDMS^{9K} on glass[h]	Water	–	–	106°	105°	1°	[58]
ix. PU—Fluorolink	Hexadecane	68°	7°	–	–	–	[59]
x. C10 sol gel hybrid film	Hexadecane	–	3.4°	36°	34°	2°	[60]
xi. PU—2 % SiClean	Water	90°	31°	104°	76°	28°	[57]
	Hexadecane	31°	6°	–	–	–	[61]
xii. PU—8 % SiClean	Water	104°	23°	106°	88°	18°	[57]
	Hexadecane	34°	2°	–	–	–	[61]

[a]Static contact angle, estimated error <2
[b]Sliding angle, estimated error <3
[c]Advancing contact angle, estimated error <2
[d]Receding contact angle, estimated error <2
[e]Contact angle hysteresis
[f]Pillar diameter/pitch/height/sidewall
[g]Square pillar/pitch/height
[h]2K and 9K denote the molecular weight of the PDMS chains

and display; anti-icing coating for power lines, rooftops, and airplanes; corrosion-resistant coating for bridges and other metal structures; drag reduction in ship; gas and fuel transportation; microfluidic devices; etc., are being pursued worldwide. Many interesting surface properties have been reported. Some of the counterintuitive observations include (1) large contact angle surfaces with large sliding angles and hysteresis and (2) small contact angle surfaces with low hysteresis and high drop mobility. While the literature data is massive, in Table 5.2 we highlight some of the key examples to illustrate the usefulness of a better understanding of basic concepts in comprehending these unexpected results.

Surfaces i–iv are rough surfaces whose surface properties and structures are well characterized. While Lotus leaf i [56], artificial superhydrophobic surfaces ii and iii (with water), and superoleophobic surfaces ii (with hexadecane) [51, 57] all exhibit the expected large contact angle with small sliding angle and low hysteresis, sticky rough surfaces with large contact angles in the cases of iii (hexadecane) and iv (water) are less common [44, 57]. The probing liquids in both cases fully wet the rough surfaces and are in the Wenzel state [43]. According to the Wenzel equation, the static contact angle should decrease if the surface material is hydrophilic or oleophilic, but the contrary is observed in the cases of iii (hexadecane) and iv (water). The observation is in violation with the Wenzel equation and is attributable to the pinning effect. Detail discussion of the pinning effect and the mechanism for producing the unexpected large contact angle has been given in Sect. 4.2.2 in this book.

Surfaces v–xii are smooth flat surfaces and they are all hydrophobic (water $\theta > 90°$). Their contact angles, sliding angles, and hysteresis are not correlated to common intuition, e.g., the water advancing contact angles θ for v and vi are larger than those of vii and viii, but their sliding angles and hysteresis are larger too! The results are consistent with the scattered plots shown in Figs. 5.1, 5.2, and 5.3. Counterintuitive results are also observed when hexadecane is used as the probing liquid. Surfaces v, vi, and ix–xii all exhibit low contact angles: θ, θ_A, and θ_R. While the low contact angles are expected due to the low surface tension of hexadecane (27.5 mN/m as compared to 72.8 mN/m for water), what's unexpected are their low ($<10°$) sliding angles and small hysteresis. These results imply that these surfaces are highly wettable with high adhesion. On the other hand, the small sliding angles suggest that these surfaces are actually nonsticky. How can that be? The high adhesion and nonsticky properties seem to contradict to each other. However, the observations can be rationalized based on the thermodynamic of the wetting and de-wetting process. Specifically, when $\theta_A \approx \theta_R$, the energy that is required to de-pin the drop is likely well compensated by the energy gained from the favorable wetting process. Using several low-hysteresis examples, McCarthy et al. [38, 58] conjectured earlier that kinetic factor, namely, low activation barrier for the de-pinning process, is the main reason for the high drop mobility. We feel that while low activation energy barrier to de-pin the receding contact line may be the effect, the real cause for the high mobility is still the favorable thermodynamic of the wetting/dewetting process. These low-hysteresis surfaces may find applications as anti-graffiti coating and easy-clean, self-clean surfaces for inkjet printheads [59, 61].

The unusual surface property exhibited by v–xii may offer new avenue for the designs of new self-cleaning surfaces of varying contact angles. It may shine new lights into the mechanism of contact angle hysteresis. For example, surfaces vii–xii all comprise low-surface-energy functional groups at the surfaces, e.g., the C10 hydrocarbon chain, oligomers of polydimethylsiloxane, and the perfluoropolyether polymer chain. These low-surface-energy functional groups are all known to be flexible, and they will migrate outward toward the air–surface interface during coating and drying. Incidentally, these low-surface-energy functional groups are also compatible with hexadecane. The favorable hexadecane–surface interactions have resulted in high wettability and low θ_A for surfaces ix–xii. During receding, the contact line recedes from the wetted area where equilibration at the interface between liquid molecules and the solid surfaces has already occurred. The receding angle is dictated by interfacial tension γ_{SL}^{re}. We further suggest that due to the compatibility between hexadecane and these low-surface-energy, flexible functional groups, materials at the liquid–solid interface also get their chance to equilibrate during advancing. As a result, the liquid–solid interfacial tension during advance, , γ_{SL}^{ad} is very similar to γ_{SL}^{re}. Consequently, θ_A is similar to θ_R according to Eqs. (5.5) and (5.6) and very low hysteresis is resulted. Additional evidence to support this concept comes from the contact angle data in surfaces v and vi. Comparison of the water and hexadecane contact angle data reveal that probing the surfaces with water always leads to larger contact angles, θ, θ_A, θ_R, and α as well as larger hysteresis as compared to those of hexadecane. There appears an inverted correlation between favorability of the liquid–solid interaction and contact angle. For example, with stronger interfacial interactions between hexadecane and the C18 hydrocarbon chain and the C8 perfluorocarbon chain in the monolayers of the OTS SAM and FOTS SAM, the contact angles, sliding angles, and hysteresis with hexadecane are found to be all smaller than those with water.

References

1. Young T (1805) An essay on the cohesion of fluids. Philos Trans R Soc Lond 95:65–87
2. Rayleigh L (1890) On the tension of water surfaces, clean and contaminated, investigated by the method of ripples. Philos Mag 30:386–400
3. Bartell FE, Wooley AD (1933) Solid–liquid-air contact angles and their dependence upon the surface condition of the solid. J Am Chem Soc 55:3518–3527
4. Bartell FE, Hatch GB (1934) Wetting characteristics of Galena. J Phys Chem 39:11–24
5. Dupre A (1869) Theorie Mechanique de la Chaleur. Gauthier-Villars, Pairs, p 369
6. Gibbs JW (1928) Trans. Connecticut Acad. 1876–1878, 3; Collected Works, vol. 1. Longmans, Green, New York
7. Shuttleworth R, Bailey GJ (1948) The spreading of a liquid over a rough surface. Discuss Faraday Soc 3:16–22
8. Morra M, Occhiello E, Garbossi F (1990) Knowledge about polymer surfaces from contact angle measurements. Adv Colloid Interface Sci 32:79–116
9. Joanny JF, de Gennes PG (1984) A model for contact angle hysteresis. J Chem Phys 81:552–561
10. Good RJ (1977) Surface free energy of solids and liquids: thermodynamics, molecular forces, and structures. J Colloid Interface Sci 59:398–418
11. Carre A, Shanahan MER (1995) Drop motion on an inclined plane and evaluation of hydrophobic treatment on glass. J Adhes 49:177–185
12. Bormashenko E, Bormashenko Y, Oleg G (2010) On the nature of the friction between nonstick droplets and solid substrates. Langmuir 26:12479–12482
13. Xiu Y, Zhu L, Hess DW, Wong CP (2008) Relationship between work of adhesion and contact angle hysteresis on superhydrophobic surfaces. J Phys Chem C 112:11403–11407
14. Lv C, Yang C, Hao P, Feng H, Zheng Q (2010) Sliding of water droplet on microstructured hydrophobic surfaces. Langmuir 26:8704–8708
15. Ko YC, Ratner BD, Hoffman AS (1981) Characterization of hydrophilic-hydrophobic polymer surfaces by contact angle measurements. J Colloid Interface Sci 82:25–37
16. Morra M, Occhiello E, Garbassi F (1989) Contact angle hysteresis on oxygen plasma treated polypropylene surfaces. J Colloid Interface Sci 132:504–508
17. Tsibouklis J, Nevell TG (2003) Ultra-low surface energy polymers. The molecular design requirements. Adv Mater 15:647–650
18. Strohmeier BR (1993) Improving the wettability of aluminum foil with oxygen plasma treatment. In: Mittal KL (ed) Contact angle, wettability and adhesion. VSP BV, Zutphen, pp 453–468

19. Tsai YC, Chou CT, Penn LS (1993) Contact angle data and adhesive performance for smooth surfaces with attached molecular chains. In: Mittal KL (ed) Contact angle, wettability and adhesion. VSP BV, Zutphen, pp 729–737
20. Murase H, Fujibayashi T (1997) Characterization of molecular interfaces in hydrophobic systems. Prog Org Coat 31:97–104
21. Rios PF, Dodiul H, Kenig S, McCarthy S, Dotan A (2007) The effect of polymer surface on the wetting and adhesion of liquid systems. J Adhes Sci Technol 3–4:227–241
22. Jao SHE, Chen JH, Shifley JD, Aslam M, Pavlisko JA (2009) Electrophotographic apparatus. US Patent 7,565,091
23. Hasebe S (2008) Numerical simulation of the toner melting behavior in fuser nip considering toner rheology. NIP24 and Digital Fabrication 2008. Final Program and Proceedings, Pittsburgh, pp 519–522
24. Gao L, McCarthy TJ (2008) Teflon is hydrophilic. Comments on definitions of hydrophobic, shear versus tensile hydrophobicity and wetting characterization. Langmuir 24:9183–9188
25. Mittal KL (ed) (1993) Contact angle, wettability and adhesion. VSP BV, Zutphen
26. Samuel B, Zhao H, Law KY (2011) Study of wetting and adhesion interactions between water and various polymer and superhydrophobic surfaces. J Phys Chem C 115:14852–14861
27. Extrand CW, Kumaga Y (1995) Liquid drops on an inclined plane. The relationship between contact angles, drop shape, and retention force. J Colloid Interface Sci 170:515–521
28. Extrand CW, Gent AN (1990) Retention of liquid drops by solid surfaces. J Colloid Interface Sci 138:431–442
29. Furmidge CGL (1962) Studies at phase interfaces I. The sliding of liquid drops on solid surfaces and a theory for spray retention. J Colloid Sci 17:309–324
30. Macdougall G, Ockrent C (1942) Surface energy relations in liquid/solid systems. 1. The adhesion of liquids to solids and a new method of determining the surface tension of liquids. Proc R Soc Lond A 180:151–173
31. Antonini C, Carmona FJ, Pierce E, Marengo M, Amirfazli A (2009) General methodology for evaluating the adhesion force of drops and bubbles on solid surfaces. Langmuir 25:6143–6154
32. Pierce E, Carmona FJ, Amirfazli A (2008) Understanding of sliding and contact angle results in tilted plate experiments. Colloids Surf A Physicochem Eng Asp 323:73–82
33. Krasovitski B, Marmur A (2005) Drops down the hill: theoretical study of limiting contact angles and the hysteresis range on a tilted plate. Langmuir 21:3881–3885
34. ElSherbini AI, Jacobi AM (2004) Liquid drops on vertical and inclined surfaces 1. An experimental study of drop geometry. J Colloid Interface Sci 273:556–565
35. Della Volpe C, Siboni S, Morra M (2002) Comments on some recent papers on interfacial tension and contact angles. Langmuir 18:1441–1444
36. Cheng DF, Urata C, Masheder B, Hozumi A (2012) A physical approach to specifically improve the mobility of alkane liquid drops. J Am Chem Soc 134:10191–10199
37. Tadmor R, Bahadur P, Leh A, N'guessan HE, Jaini R, Dang L (2009) Measurement of lateral adhesion forces at the interface between a liquid drop and a substrate. Phys Rev Lett 103:266101
38. Gao L, McCarthy TJ (2006) Contact angle hysteresis explained. Langmuir 22:6234–6237
39. McHale G, Shirtcliffe NJ, Newton MI (2004) Contact angle hysteresis on superhydrophobic surfaces. Langmuir 20:10146–10149
40. Pease DC (1945) The significance of the contact angle in relation to the solid surface. J Phys Chem 49:107–110
41. Johnson RE, Dettre RH (1964) Contact angle hysteresis 1. Study of an idealized rough surface. In: Fowkes F (ed) Contact angle, wettability, and adhesion, advances in chemistry. American Chemical Society, Washington, DC, pp 112–135
42. Neumann AW, Good RJ (1972) Thermodynamic of contact angles 1. Heterogeneous solid surfaces. J Colloid Interface Sci 38:341–358

43. Wenzel RN (1936) Resistance of solid surfaces to wetting by water. Ind Eng Chem 28:988–994
44. Forsberg PSH, Priest C, Brinkmann M, Sedev R, Ralston J (2010) Contact line pinning on microstructured surfaces for liquids in the Wenzel state. Langmuir 26:860–865
45. Extrand CW (2002) Model for contact angles and hysteresis on rough and ultraphobic surfaces. Langmuir 18:7991–7999
46. Bartell FE, Bjorklund CW (1952) Hysteresis of contact angles. A study of interfacial contact angles in the mercury-benzene-water system. J Phys Chem 56:453–457
47. Chen YL, Helm CA, Israelachville JN (1991) Molecular mechanisms associated with adhesion and contact angle hysteresis of monolayer surfaces. J Phys Chem 95:10736–10747
48. Holly FJ, Refojo MF (1975) Wettability of hydrogels 1. Poly(2-hydroxyethyl methacrylate). J Biomed Mater Res 9:315–326
49. Lee S, Park JS, Lee TR (2008) The wettability of fluoropolymer surfaces: influence of surface dipoles. Langmuir 24:4817–4826
50. Extrand CW, Kumagai Y (1997) An experimental study of contact angle hysteresis. J Colloid Interface Sci 191:378–383
51. Zhao H, Park CK, Law KY (2012) Effect of surface texturing on superoleophobicity, contact angle hysteresis and "robustness". Langmuir 28:14925–14934
52. Nosonovsky M (2007) Model for solid–liquid and solid–solid friction of rough surfaces with adhesion hysteresis. J Chem Phys 126:224701
53. Kanungo M, Mettu S, Law KY, Daniel S (2014) Effect of roughness geometry on wetting and de-wetting of PDMS surfaces. Langmuir 30:7358–7368
54. Law JBK, Ng AMH, He AY, Low HY (2014) Bioinspired ultrahigh water pinning nanostructures. Langmuir 30:325–331
55. Mettu S, Kanungo M, Law KY (2013) Anomalous thermally induced pinning of a liquid drop on a solid substrate. Langmuir 29:10665–10673
56. Koch K, Barehlott W (2009) Superhydrophobic and superhydrophilic plant surfaces: an inspiration for biomimetic materials. Philos Trans R Soc A Math Phys Eng Sci 367:1487–1509
57. Zhao H, Law KY, Sambhy V (2011) Fabrication, surface properties, and origin of superoleophobicity for a model textured surface. Langmuir 27:5927–5935
58. Krumpfer JW, McCarthy TJ (2010) Contact angle hysteresis. A different view and a trivial recipe for low hysteresis hydrophobic surfaces. Faraday Discuss 146:103–111
59. Sambhy V, Law KY, Zhao H, Chugh S (2013) Thermally stable oleophobic low adhesion coating for inkjet printing front face. US Patent 8,544,987
60. Urata C, Cheng DF, Masheder B, Hozumi A (2012) Smooth, transparent and non-perfluorinated surfaces exhibiting unusual contact angle behavior towards organic liquids. RSC Adv 2:9805–9808
61. Kovacs GJ, Law KY, Zhao H, Sambhy V (2012) Coating for an ink jet printhead front face. U.S. Patent 8,226,207

Chapter 6
Terminologies and Definitions

Abstract One of the weak links in surface research has been the lack of clear, well-defined terminologies, definitions, and common language. This at least in part contributes to the current messiness in the surface literature. In this chapter, some of the commonly used terminologies and language are overviewed, shortfalls are discussed, and areas for improvement are proposed. In terms of definition, hydrophilicity and hydrophobicity are the most important concepts in surface science, and they mean having and lacking of affinity with water, respectively. Water contact angle θ at 90° has been defined as the cutoff between hydrophilicity and hydrophobicity. This definition may have been derived from trigonometry but has been challenged numerous times in the past. An improved definition based on the wetting and adhesion interactions of water with 20 different surfaces of varying water affinity is proposed. Surfaces with $\theta_R > 90°$ were found to have no affinity with water and are defined as hydrophobic, whereas those with $\theta_R < 90°$ are defined as hydrophilic. Surfaces with $\theta_A \geq 145°$ are shown to have no attraction toward water. Accordingly, a surface can be defined as superhydrophobic when its θ_R is $>90°$ and θ_A is $\geq 145°$. The fundamental reason why a surface becomes hydrophobic is discussed. The methodology has been extended to define hexadecane oleophilicity, oleophobicity, and superoleophobicity. It is also shown that the philicity/phobicity cutoff should not be universal at 90°, rather it should be dependent of the liquid surface tension; the lower the surface tension, the larger the θ_R cutoff angle.

Keywords Hydrophilicity • Hydrophobicity • Definitions • Terminologies • Oleophilicity • Oleophobicity • Superhydrophobicity • Superoleophobicity • Liquid cohesion • Wetting interaction • Amphiphobicity • Omniphobicity

6.1 Background

Surface is a very important branch of physical science and has been studied mostly by chemists and physicists since the nineteenth century. Technically, when two or more objects come together, the interfaces between them will be part of surface science. Very often, interfacial interactions are more dominant and interesting than bulk material properties. As a result, surface has become an intersection point of many fields of research, and the studies have attracted a very diverse group of researchers and practitioners. Nowadays, in addition to chemists and physicists,

K.-Y. Law, H. Zhao, *Surface Wetting*, DOI 10.1007/978-3-319-25214-8_6

materials scientists, engineers of nearly all disciplines (materials, chemical, civil, mechanical, aerospace, electric, etc.), biologists, and manufacturers are also studying surfaces because of its relevance to the subject matter of their interests. On one hand, this is good for the science as it will be benefitted from the very diverse participation. What's unfortunate is that there is no universal, technical language in surface science to allow the community to converse effectively. Quite often, definitions are improperly defined, terminologies are created arbitrarily, and expressions sound like local dialects. This has led to misunderstanding, confusions, and sometimes ad hoc phrases. In this chapter, some commonly used terminologies, descriptors, and definitions in surface are overviewed. Assessment, constructive criticism, and areas for improvement are proposed based on the fundamental of wetting, namely, the movement of the contact line. Our goal is to create a conversation, which may catalyze the different segments of researchers and practitioners, to work toward a set of common languages, terminologies, and definitions. It is also our hope that this work will lay the framework for "the authority" or leaders in this field to officially put together some standardized terminologies and definitions for the community at large.

6.2   Common Terminologies and Languages

6.2.1   Terminologies

The first information one seeks when a liquid wets a solid surface is wettability. Will the liquid wet the surface or repel from it? Many terminologies describing the wetting process are originated from Greek or Latin words. Usually, the terms comprise two parts. The first part describes the type of liquid, and the second part reveals its affinity. The word "philic" means with affinity in Greek, and "phobic" is the opposite, which means lack of affinity. A summary of some of the common terminologies along with their descriptions from an English dictionary is outlined in Table 6.1. For simplicity, the second half of the terms (phobicity) are not shown in the table.

Water is the most tested liquid in surface science because of its abundance. Its Greek word is hydro. Therefore, hydrophilic and hydrophobic are the most

Table 6.1 Common terminologies used in the surface literature

Term	(Original) description	Remarks
Hydrophilic	Affinity with water	
Oleophilic	Affinity with oils	
Amphiphilic	Surfactant molecule having a polar group and a hydrocarbon chain	Chemists have extended it to describe polar and nonpolar interactions
Lyophilic	Affinity with colloids	Not appropriate to describe liquid–solid interaction
Lipophilic	Affinity with fats and lipids	Limit in scope
Omniphilic	Affinity with every liquid	

recognized terms in the surface literature. Other terms that we thought are appropriate to describe liquid–solid interactions are oleophilic/oleophobic and omniphilic/omniphobic. Each pair of terminology actually includes groups or classes of liquids. Ole means oil and is a class of liquids that are immiscible with water. It is not specific and covers hydrocarbon oils, oils extracted from natural products as well as synthetic oils. Omni means every liquid, covering liquids from polar to nonpolar, from water to alcohols, ketones, esters, amides, halocarbons, hydrocarbons, and so on. To be more informative, authors are advised to specify their liquid(s) when these terms are used in their publications.

The term amphiphilic is used to describe surfactant-like molecules having a dual polarity: a polar group in one end and a long hydrocarbon chain in the other end. Chemists have stretched it to describe polar and nonpolar interactions. If the term amphiphilic/amphiphobic is used, the author is advised to specify the liquids used. As for lipophilic, we find it limits in scope because lipo usually means fats and lipids. Although some have extended it to hydrocarbon oils, we feel that hydrocarbon oils are already represented by oleo. We also feel that the term lyophilic is not appropriate. It was used to describe affinity between colloids and their dispersed media and doesn't appear to be too relevant.

Although the terminologies in Table 6.1 have not been a source of serious confusion, it does not hurt if we sharpen up with our language to effectively communicate our results. For instance, everyone knows that hydro means water. Can the community agree that oleo means hexadecane? As for "omni" and "amiphi," it would make sense for the authors to study a set of liquids with varying surface tension, viscosity, functionality, and polarity to substantiate the claim. In addition to water and hexadecane (or decane), the liquid set should include alcohol (ethylene glycol), ketone (methyl ethyl ketone), ester (ethyl acetate), amide (dimethylformamide), and halocarbon (diiodomethane), for example. If the test liquids for each terminology are well defined, this will leave little guesswork to the reader and inappropriate claims by the authors.

6.2.2   Languages and Descriptors

Figure 6.1 schematically depicts the three interactions between a liquid droplet and a surface. These three interactions are actually governed by the movement of the contact line. When the liquid first wets the surface, the contact line advances outward, and the first information one seeks is wettability. The adjectives to describe the surface are wettable and non-wettable. As for the liquid, it will either wet or partially wet the surface or repel from it. As discussed in Chap. 5, wettability is measured by the advancing angle θ_A. Once the liquid partially wets the surface, a static sessile drop is formed. There exist two interactions between the sessile drop and the surface. In the vertical direction, it is the adhesion and it is measured by the receding angle θ_R. The only motion for the contact line is receding, and an interface (liquid–solid) is eliminated when the liquid droplet is detached from the surface.

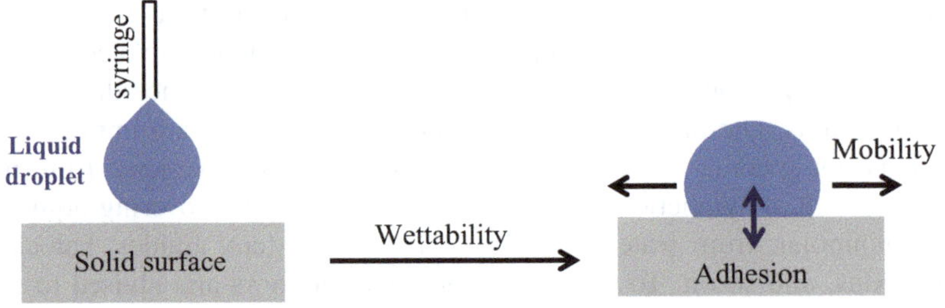

Fig. 6.1 Summary of interactions when a liquid drop contracts a solid surface

Table 6.2 Suggested descriptors for liquid–solid interactions

Interaction	Measured by	Descriptors/adjectives for		Moving contact line
		Surface	Liquid	
Wettability	θ_A	Wettable/non-wettable	Wetting/repelling	Advancing
Adhesion	θ_R	High/low adhesion	Sticky/nonsticky	Receding
Horizontal mobility	α or $(\cos\theta_R - \cos\theta_A)$	Sticky/slippery	Mobile/immobile	Advancing and receding

In the case of moving a horizontal drop, both advancing and receding contact lines are in action, and the wetting state remains the same as the liquid is moving along the surface. If the drop is stuck and has no lateral mobility, one can simply describe the surface as sticky and the drop as immobile. The mobility of the drop is measured by the sliding angle α, which is proportional to $(\cos\theta_R - \cos\theta_A)$. Gao and McCarthy [1] used the terms tensile and shear hydrophobicity to describe the vertical and horizontal interactions. The proposal is faulty and inappropriate. For example, the use of the term shear hydrophobicity to describe lateral mobility is technically incorrect as the wettability of the drop (contact angle) remains unchanged during the movement. As pointed out in Chap. 5 (Table 5.2), highly non-wettable surfaces can exhibit adhesion ranging from sticky to slippery. Similarly, slippery surfaces can have a range of wettability too! We suggest differentiating these two interactions through the movement of the contact line. For adhesion, the only movement for the contact line is receding, and an interface (liquid–solid) is eliminated when the liquid droplet is detached from the surface. For lateral movement, both advancing and receding contact lines continue to move, and drop affinity to the surface remains the same. A summary for these three interactions, their measurement methods, the recommended descriptors, and the movement of the contact line are summarized in Table 6.2.

The shortfall in our vocabulary is exposed when describing highly mobile drops with contact angles ranging from ~30 to ~120°. Three groups of researchers, the McCarthy group at the University of Massachusetts [2, 3], the Hozumi group at AIST (Advanced Industrial Science and Technology) Japan [4–6], and the Law group at Xerox [7–9], recently reported that liquid drops can have high mobility on wettable and even highly wettable surfaces. Some of the key data are summarized in Table 6.3.

Table 6.3 Key contact angle data for recently reported surfaces with high liquid mobility

Surface	Liquid	θ	θ_A/θ_R	α	References
Alkyl silane from $Cl(SiMe_2O)_n SiMeCl$	Water		~105°/~104°		[2]
	CH_2I_2		~74°/~72°		
	Hexadecane		~36°/~35°		
Alkyl silane from (Me_3SiO) $SiCH_2CH_2Si(CH_3)_2Cl$	Water		~104°/~103°		[2]
	CH_2I_2		~71°/~66°		
	Hexadecane		~38°/~36°		
Grafted PDMS^{2K} on glass[a,b]	Water		~105°/~104°		[3]
Grafted PDMS^{6K} brush on wafer[a]	Water		~108°/~104°		[5]
	Hexadecane		~35°/~35°		
C10 sol–gel hybrid film	Water		~109°/~100°	40°	[6]
	CH_2I_2		~73°/~71°	6.2°	
	Hexadecane		~36°/~34°	3.4°	
PU-8 % Silclean[c]	Hexadecane	~34°		3.4°	[7]
PU-2 % Fluorolink[d]	Hexadecane	~68°		7°	[8]

[a]2K and 6K denote the molecular weight of the PDMS chains
[b]Low hysteresis were observed with CH_2I_2 and hexadecane
[c]Silclean is a polyhydroxy PDMS oligomer, which is incorporated into the polyurethane polymer during cross-linking
[d]Fluorolink is a dihydroxy-terminated perfluoropolyether, which is incorporated into the polyurethane backbone during cross-linking

Other than with water, the contact angles of most of the surfaces are <90° with the tested liquids, indicating that they are all highly wettable. McCarthy et al. [2] used the terms ultrahydrophobicity and ultralyophobicity to describe the unexpected mobility for water, diiodomethane, and hexadecane on their silanated silicon wafer surfaces. Hozumi et al. [6] described the same high-mobility phenomenon as dynamic de-wettability on their study of the surface properties of alkyl-terminated hybrid sol–gel films. We feel that none of these terminologies are technically correctly describing the observation. Phobicity is a recognized descriptor for lack of affinity. The terms ultrahydrophobicity and ultralyophobicity generally stand for super repellency than for the high mobility. Moreover, a moving droplet has nothing to do with wettability or de-wettability. During drop moving, the contact area underneath the droplet is constant, so neither wetting nor de-wetting occurs.

It is important to point out that it is not against any fundamental law of chemistry or physics to have a wettable surface exhibiting low hysteresis and small sliding angle at the same time. In fact, Della Volpe, Siboni, and Morra [10] noted in their comments to Langmuir that hysteresis and sliding angle do not have to be dependent of surface hydrophobicity. In Zisman's 1964 article [11], he cited the collaborative work between Langmuir and Blodgett (LB), who observed that white mineral oil was able to roll off a trimolecular stearate LB film on glass, which exhibited a contact angle of ~55°. Langmuir and Blodgett attributed the unusual "de-wetting" property to the tight packing of the C18 chain in the LB film as well as the noninteracting nature of the CH_3 end group [12]. In 1996, Schmidt et al. [13] reported the synthesis of a new family of nonsticking, wettable polymers by cross-linking reactive perfluoroalkyl polymeric surfactants with poly(2-isopropenyl-2-oxazoline) at different reactant ratios under a various reacting and curing conditions. Static, advancing/receding angles and sliding angles with water and hexadecane were reported. With hexadecane, four of the polymer coatings exhibit contact angles range between 58 and 67° with sliding angles less than 15°. These polymer coatings are generally hydrophobic with water contact angles range between 101 and 112° along with slightly larger sliding angles, 24–27°. Thus, wettable, nonsticking surfaces are not new. We assume that the creation of new terms may have been influenced by the hype of superhydrophobicity in recent years. For simplicity and clarity, we don't see the need of any new descriptors. Simple adjectives, such as sticky and slippery, are sufficient to describe these surfaces; likewise, the terms mobile and immobile should be adequate in describing drop mobility. Moreover, there are always adverbs like super or highly to further modifying the adjectives.

6.3   Definitions

6.3.1   *Hydrophilicity, Hydrophobicity, and Superhydrophobicity*

After a terminology is defined and quantified, it becomes a concept. Hydrophobicity and hydrophilicity are among the most important concepts in surface science. In the scientific community, we generally accept that a surface is hydrophobic when its static water contact angle θ is >90° and is hydrophilic when θ is <90°. Actually, there is little technical rationalization or understanding why a surface suddenly becomes hydrophobic when θ just increases 1° from the cutoff point. What is the mechanism or molecular origin that leads to this change of surface property? In fact, this definition was questioned by Gao and McCarthy who called "Teflon hydrophilic" because of its high adhesion with water [1]. Others have also noticed this shortfall in the definition. van Oss [14] proposed to use the free energy of hydration (ΔG_{sl}) as the measure of hydrophilicity and hydrophobicity. Based on the analysis of the free energy of hydration for a number of compounds, he found that hydrophobic compounds attract each other in water when $\Delta G_{sl} > -113$ mJ/m^2 and that they repel each other when $\Delta G_{sl} < -113$ mJ/m^2. The value -113 mJ/m^2 was proposed to be the cut off between hydrophobicity and hydrophilicity. Vogler [15] proposed a cutoff of $\theta \sim 65°$ based on the appearance and disappearance of long-range hydrophobic interactions in his surface force measurement. The latter two definitions actually have little to do with water–surface affinity. Rather, the cutoff point shows the raise of the hydrophobic effect of surfactant molecules or protein in water. Therefore, there is a definite need of a good definition that is supported by technical data.

In 2011, Samuel, Zhao, and Law [16] reported a systematic study of the wetting and adhesion interactions of water with 20 different surfaces (**1–20**) using the microbalance in a tensiometer. Details of the study have been provided in Sect. 5.2. Briefly, we found that surface adhesion decreases linearly as the receding angle

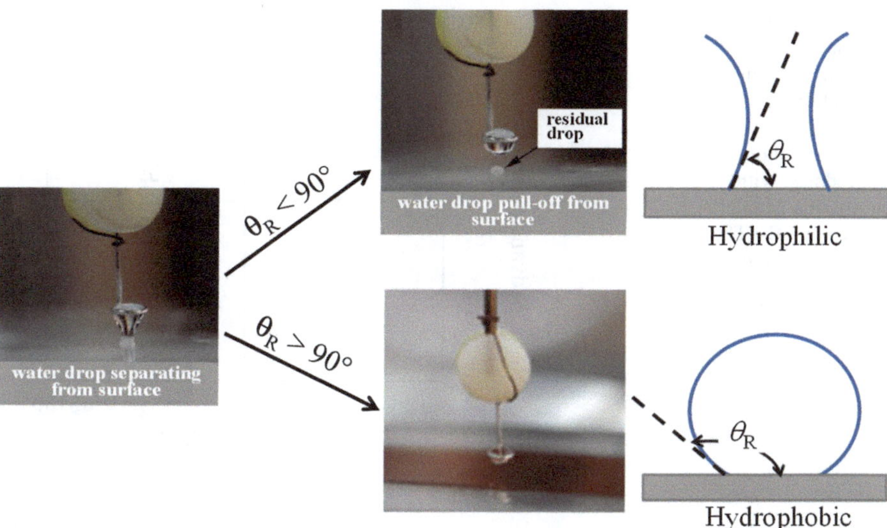

Fig. 6.2 Photographs and schematics showing water–surface separation from a hydrophilic (*top*) and hydrophobic (*bottom*) surface (reproduced with permission from [17], copyright 2014 The American Chemical Society)

increases (Fig. 5.9). As pointed out in Sect. 5.2.4, we also notice a clear cutoff at $\theta_R \sim 90°$. Specifically, for surfaces with $\theta_R < 90°$, a tiny residual water droplet was observed after the water droplet is separated from the surface. The surface is clean for surfaces with $\theta_R > 90°$. Photographs and the schematic showing these two cases are given in Fig. 6.2. The observation of the tiny residual water droplet after the water droplet and the surface separate is indicative of definite affinity. The clear distinction between surfaces at $\theta_R < 90°$ and $\theta_R > 90°$ leads to the proposal that a surface is defined as hydrophilic when its water θ_R is $<90°$ and is hydrophobic when θ_R is $>90°$ [17].

Another intriguing question is what is the fundamental driver to make the surface changes from hydrophilic to hydrophobic at the cutoff $\theta_R \sim 90°$. In Chap. 5, we also show that there exists a good correlation between the wetting force and advancing contact angle θ_A (Fig. 5.5). Attractive interaction is shown to decrease as θ_A increases. The result indicates that there is always positive water–surface attraction and that this attraction is weakening as the surface becomes more hydrophobic. The fact that no residual water droplet was observed when $\theta_R > 90°$ can be attributed to the high cohesion of the water droplet. In other words, the water droplet prefers to be in the droplet state than wetting the surface due to the small wetting energy. The importance of liquid cohesion in wetting was in fact discussed lengthy in Young's original essay [18].

Another term that has been tentatively defined in the literature is superhydrophobicity. As noted by Roach, Shirtcliffe, and Newton [19], the definition of super-

Fig. 6.3 A 2D graphic representation for the definitions of surface hydrophilicity, hydrophobicity, and superhydrophobicity (reproduced with permission from [17], copyright 2014 The American Chemical Society)

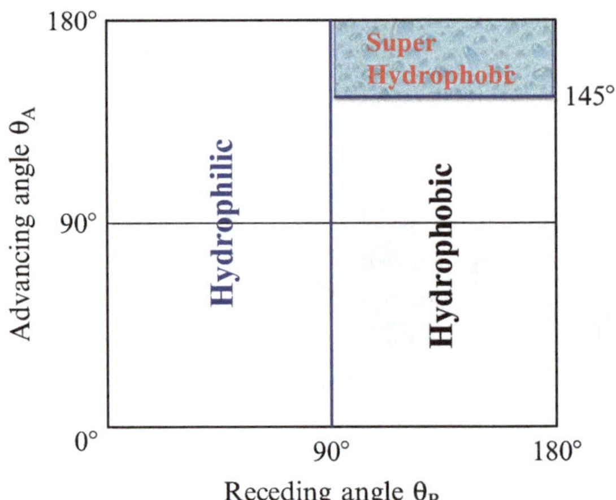

hydrophobicity for surfaces having water $\theta > 150°$ is somewhat arbitrary. On the other hand, the result in Fig. 5.5 actually shows that there is practically no wetting attraction between water and surfaces with $\theta_A \geq 145°$. Hence, a hydrophobic surface can be moved to the super status when there is absolutely no wetting interaction with water. A 2D representation of the surface definition for superhydrophobicity is given in Fig. 6.3.

In recent years, the term sticky superhydrophobicity, which is a convenient description of a sticky drop that has a large contact angle, appeared. Some Wenzel droplets also exhibit very large contact angle [20]. With the definition described in this chapter, the ad hoc term sticky superhydrophobicity should no longer be in existence. Such surface will be simply hydrophilic as its receding angle is <90°.

6.3.2 Oleophilicity, Oleophobicity, and Superoleophobicity

Surface wettability is known to increase as the surface tension of the wetting liquid decreases [11]. As a result, considerable amount of attention has been paid recently to surfaces which are non-wettable against hydrocarbon oil such as hexadecane [21–25]. It is generally believed that a surface that is super repellent against heptane to hexadecane should repel most liquids due to their low surface tensions. In the literature, the traditional cutoff for water has been extended to hexadecane, namely, a surface is oleophilic if its hexadecane θ is <90°, oleophobic if θ is >90°, and superoleophobicity if θ is ≥150°. To our knowledge, there is no technical justification for this definition. Law [17, 26] recently examined some preliminary data on the wetting and adhesion interactions between hexadecane and several surfaces using the microbalance technique as described in Chap. 5. The preliminary results are summarized in Fig. 6.4. Results in Fig. 6.4a show that the wetting force decreases

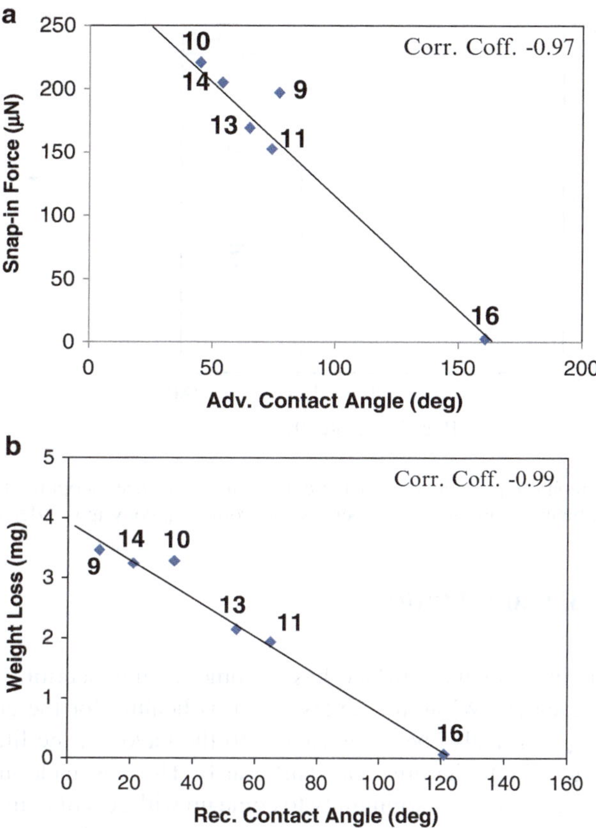

Fig. 6.4 Plots of (**a**) snap-in force versus advancing contact angle and (**b**) weight loss of the hexadecane droplet after pull off versus receding contact angle with different surfaces (reproduced with permission from [26], copyright 2015, De Gruyter)

linearly as the hexadecane θ_A on the surface increases. An intercept at 163° is observed, indicating that surfaces with $\theta_A \geq 163°$ will have no affinity with hexadecane. This trend is similar to that observed with water (Fig. 5.5). The plot between adhesion force and θ_R is more scattered and is attributable to the high adhesion between hexadecane and the probing surfaces. A significant quantity of hexadecane was found transferring to the surface after pull off, which complicate the force measurement. Since the amount of hexadecane that is transferred is expected to increase as the surface adhesion increases, the trend can then be captured as the weight loss during the pull-off experiment. Figure 6.4b shows the relationship between the weight loss of the hexadecane droplet after pull off and surface θ_R. A much better plot is obtained. The intercept is at 124°, suggesting that surfaces with $\theta_R > 124°$ should have little affinity with hexadecane. Accordingly, the cutoff between oleophilicity/phobicity for hexadecane is at θ_R 124°, and a surface can be defined as superoleophobic when θ_R is >124° and θ_A is ≥163°. The 2D surface definition plot for hexadecane is given in Fig. 6.5.

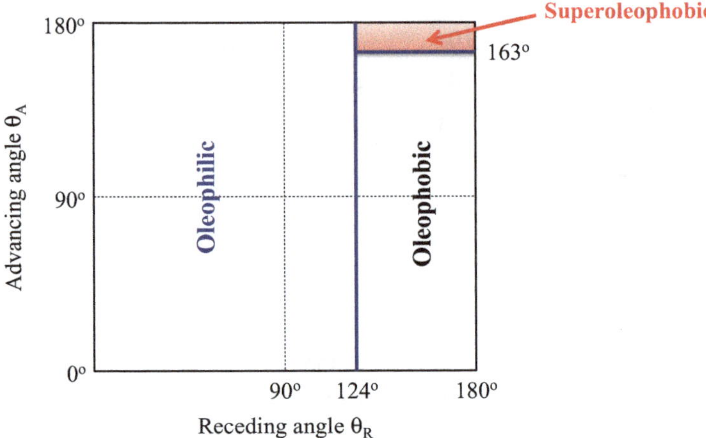

Fig. 6.5 A 2D graphic representation for the definitions of surface oleophilicity, oleophobicity, and superoleophobicity (reproduced with permission from [26], copyright 2015, De Gruyter)

6.4   Summary and Outlook

This chapter points out that surface has become an intersection point of many disciplines of research. While the diverse input is healthy for the development of surface science, it may also be a contributor to the havoc in the literature, where definitions and terminologies are created arbitrarily. This has made surface research messy, and we are urging the community to come up with sets of common language, terminology, and definition so that the communication can be more effective and less confusing.

This chapter also summarizes two new and improved definitions, one on hydrophilicity and hydrophobicity and the other on oleophilicity and oleophobicity. While the proposed 2D definitions represent an improvement over the existent ones, even more important is that it provides a fundamental reason why a surface changes from philic to phobic against a given liquid. The switching from philicity to phobicity is shown to be a result of the larger cohesion force within the droplet relative to the wetting force. Additionally, the study further reveals that philicity/phobicity cutoff should not be universally at 90° for all liquids, the cutoff should be a function of the surface tension, the lower the surface tension, the larger the θ_R cutoff angle.

References

1. Gao L, McCarthy TJ (2008) Teflon is hydrophilic. Comments on definitions of hydrophobic, shear versus tensile hydrophobicity and wetting characterization. Langmuir 24:9183–9188
2. Chen W, Fadeev AY, Hsieh MC, Oner D, Youngblood J, McCarthy TJ (1999) Ultrahydrophobic and ultralyophobic surfaces: some comments and examples. Langmuir 15:3395–3399

3. Krumpfer JW, McCarthy TJ (2010) Contact angle hysteresis: a different view and a trivial recipe for low hysteresis hydrophobic surfaces. Faraday Discuss 146:103–111

4. Cheng DF, Urata C, Masheder B, Hozumi A (2012) A physical approach to specifically improve the mobility of alkane liquid drops. J Am Chem Soc 134:10191–10199

5. Urata C, Cheng DF, Masheder B, Hozumi A (2012) Smooth, transparent and non-perfluorinated surfaces exhibiting unusual contact angle behavior towards organic liquids. RSC Adv 2:9805–9808

6. Urata C, Masheder B, Cheng DF, Miranda DF, Dunderdale GJ, Miyamae T, Hozumi A (2014) Why can organic liquids move easily on smooth alky-terminated surfaces? Langmuir 30:4049–4255

7. Kovacs GJ, Law KY, Zhao H, Sambhy V (2012) Coating for an ink jet printhead front face. US Patent 8,226,207

8. Sambhy V, Law KY, Zhao H, Chugh S (2013) Thermally stable oleophobic low adhesion coating for inkjet printhead front face. US Patent 8,544,987

9. Sambhy V, Law KY, Zhao H, Chugh S (2014) Low adhesion sol gel coatings with high thermal stability for easy clean, self cleaning printhead front face applications. US Patent 8,851,163

10. Della Volpe C, Siboni S, Morra M (2002) Comments on some recent papers on interfacial tension and contact angles. Langmuir 18:1441–1444

11. Zisman WA (1964) Relation of the equilibrium contact angle to liquid and solid constitution. In: Fowkes F (ed) Contact angle, wettability, and adhesion, advances in chemistry. American Chemical Society, Washington, DC, pp 1–51

12. Langmuir I (1934) Mechanical properties of monomolecular films. J Franklin Inst 218:143–171

13. Schmidt DL, DeKoven BM, Coburn CE, Potter GE, Meyers GF, Fischer DA (1996) Characterization of a new family of nonwettable nonstick surfaces. Langmuir 12:518–529

14. van Oss CJ (1994) Interfacial forces in aqueous media. Marcel Dekker, New York

15. Vogler EA (1998) Structure and reactivity of water at biomaterial surfaces. Adv Colloid Interface Sci 74:69–117

16. Samuel B, Zhao H, Law KY (2011) Study of wetting and adhesion interactions between water and various polymer and superhydrophobic surfaces. J Phys Chem C 115:14852–14861

17. Law KY (2014) Definitions for hydrophilicity, hydrophobicity and superhydrophobicity getting the basics right. J Phys Chem Lett 5:686–688

18. Young T (1805) An essay on the cohesion of fluids. Philos Trans R Soc London 95:65–87

19. Roach P, Shirtcliffe NJ, Newton MI (2008) Progress in superhydrophobic surface development. Soft Matter 4:224–240

20. Forsberg PSH, Priest C, Brinkmann M, Sedev R, Ralston J (2010) Contact line pinning on microstructured surfaces for liquids in the Wenzel states. Langmuir 26:860–865

21. Law KY (2014) Superoleophobic surfaces. Surface properties, fabrication methods, and potential application. In: Lyshevski S (ed) Dekker encyclopedia of nanoscience and nanotechnology, 3rd edn. CRC Press, New York, p 4657

22. Bellanger H, Darmanin T, deGivenchy ET, Guittard F (2014) Chemical and physical pathways for the preparation of superoleophobic surfaces and related wetting theories. Chem Rev 114:2694–2716

23. Kota AK, Mabry JM, Tuteja A (2013) Superoleophobic surface design: design criteria and recent studies. Surf Innov 1:77–89

24. Liu K, Tian Y, Jiang L (2013) Bio-inspired superoleophobic and smart materials: design, fabrication, and application. Prog Mater Sci 58:503–564

25. Nishimoto S, Bhushan B (2013) Bioinspired self-cleaning surfaces with superhydrophobicity, superoleophobicity, and superhydrophilicity. RSC Adv 3:671–690

26. Law KY (2015) Water-surface interactions and definitions for hydrophilicity, hydrophobicity and superhydrophobicity. Pure Appl Chem 87(8):759–765

Chapter 7
Determination of Solid Surface Tension by Contact Angle

Abstract In this chapter, approaches to determine solid surface tension by contact angle are briefly reviewed and assessed. These approaches include the Zisman method, various versions of the surface tension component methods, and the equation of state methods. The Zisman method is an empirical approach based on the relationship between the cosine of the contact angle and the surface tensions of the test liquids. The approach allows the determination of the critical surface tension of the solid. However, it is limited to low surface energy surfaces as data points from high surface tension liquids deviate from linearity due to polar and H-bonding interactions. The surface tension component approach is pioneered by Fowkes who assumed that (1) surface tension can be partitioned into individual independent components and (2) the work of adhesion can be expressed as the geometric means of the surface tension components. The original Fowkes method only considered dispersion interaction, and the methodology has been extended to include polar and H-bonding interactions in the extended Fowkes method or electron donor and acceptor interactions in the vOCG method. The equation of state assumes that the interfacial liquid–solid surface tension depends on the surface tension of the liquid and solid only. The interface surface tension was obtained by curve fitting with contact angle data and adjustable parameters. While the equation of state approach has been improved and three different versions have been developed, the basic thermodynamic assumption and the methodology were seriously challenged by many researchers in the field. It is important to note that both surface tension component methods and equation of state methods are semiempirical and that there are many assumptions in each methodology. Both approaches inherit a reversible work-of-adhesion assumption from Dupre. Specifically, for two immiscible liquids, the free energy change at the interface is equated to the interfacial tension of the newly formed interface subtracted by the surface tensions of the precursor liquids. The validity of this assumption is always questionable when one of the components is solid as the surface molecules or segments in solid have no mobility during any interfacial interaction. In view of this questionable assumption and the semiempirical nature of the contact angle approach, we propose a simpler and more direct approach to move forward. Since the motivation of determining surface tension is to be able to predict surface wettability and adhesion, we suggest measuring the advancing and receding angle of the solid surface instead. They have recently been shown to correlate to wettability and adhesion, respectively, by force measurements.

© Springer International Publishing Switzerland 2016

K.-Y. Law, H. Zhao, *Surface Wetting*, DOI 10.1007/978-3-319-25214-8_7

Keywords Solid surface tension • Solid surface energy • Contact angle • Work of adhesion • Zisman method • Surface tension component method • Fowkes method • Owens–Wendt–Rabel–Kaelble method • Extended Fowkes method • Equation of state

7.1   Introduction

Solid surface energy is one of the basic physical properties for solid like density, melting point, refractive index, dielectric constant, modulus, etc. It is a very important material property for anyone, who is interested in interfacial interaction between a solid and another material. The industrial needs are very diverse. While the obvious applications are paints, coatings, and adhesives, the less obvious ones actually have much higher economic values to date. They include cleaning in the semiconductor industry, printing in printed electronic device manufacturing, aerospace, shipping, mining, pharmaceuticals, and so on. Generally speaking, a high surface energy material will be more reactive and stickier and vice versa. Such property is thought to originate from interactions at the atomic and molecular level and has not been accessible for direct measurement [1]. About 200 years ago, Young [2] studied liquid wetting on solid surface and found that an angle is formed at the three-phase contact line as a result of the mechanical equilibrium between the liquid surface tension (γ_{LV}), solid surface tension (γ_{SV}), and liquid–solid interfacial tension (γ_{SL}). His description has become the famous Young's equation (Eq. 7.1), and the subject was detailed in Chap. 3 in this book.

$$\gamma_{SV} = \gamma_{LV} \cdot \cos\theta + \gamma_{SL} \tag{7.1}$$

The Young's equation is problematic because only two out of four quantities (γ_{LV} and θ) can be determined experimentally. Dupre [3] later introduced the concepts of reversible work of cohesion (W^{coh}) and work of adhesion (W^{ad}) between two liquids. A concept diagram is depicted in Fig. 7.1. Hypothetically, when two cylinders of the same liquid are brought together (Fig. 7.1a), the free energy change per unit area ($\Delta G_{11}{}^{coh}$) is the free energy of cohesion and is equaled to the negative of work of cohesion and can be expressed as

Fig. 7.1 Diagram illustration of the concepts of reversible work of (**a**) cohesion and (**b**) adhesion

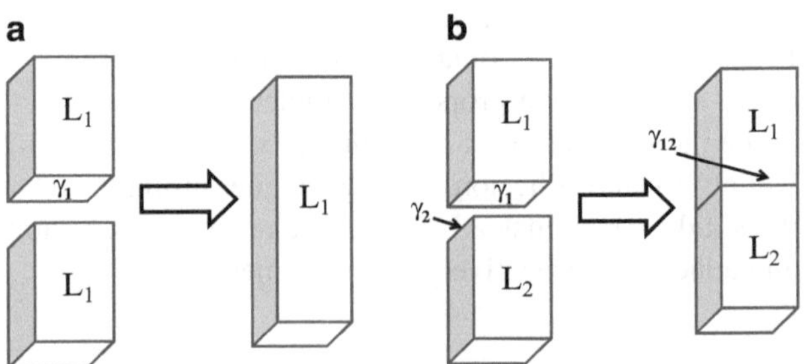

$$\Delta G_{11}^{coh} = -2\gamma_1 = -W_{11}^{coh} \tag{7.2}$$

where γ_1 and W_{11}^{coh} are the surface tension and work of cohesion for L_1, respectively.

Similarly, when two cylinders of immiscible liquids are brought together (Fig. 7.1b), the free energy change per unit area would be ΔG_{12}^{ad} and is equaled to the negative of work of adhesion. More importantly, the free energy change can be equated as the interfacial tension of the newly formed interface subtracted by the surface tensions of the precursor liquids.

$$\Delta G_{12}^{ad} = \gamma_{12} - \gamma_1 - \gamma_2 = -W_{12}^{ad} \tag{7.3}$$

where γ_2 is the surface tension of L_2 and γ_{12} is the interfacial tension between L_1 and L_2.

Now if one assumes that L_1 is a liquid and L_2 is a solid (S) and combines Eqs. (7.1) and (7.3), the work of adhesion between a liquid and a solid surface (W_{SL}^{ad}) is given by Eq. (7.4), which becomes known as the Young–Dupre equation [4]:

$$W_{SL}^{ad} = \gamma_{LV}\left(1 + \cos\theta\right) \tag{7.4}$$

This derivation opens the possibility of studying γ_{SL} through W_{SL}^{ad}, which can now be determined by measuring θ with a known liquid. Solid surface tension γ_{SV} is no longer inaccessible. Many theories and models had been studied to develop the expressions for work of adhesion W_{SL} and the corresponding interfacial tension γ_{SL}. One of the objectives of this chapter is to provide a brief overview of some key approaches for the estimation of solid surface tension through contact angle measurements. The different approaches are assessed and discussed. It is important to point out that solid surface energy and surface tension have been used interchangeably in recent literature. They are actually known to be two distinct quantities in solid [5–7], and the quantity derived from contact angle is solid surface tension. The fundamental issues around surface tension versus surface energy and the methodology of using contact angle to determine solid surface tension will be discussed, and a path forward is proposed.

7.2   Approaches to Determine Solid Surface Tension by Contact Angle

7.2.1   Zisman Method

In 1950, Fox and Zisman [8] reported their study of the wettability of poly(tetrafluoroethylene) (PTFE) with n-alkanes, di(n-alkyl)ethers, n-alkylbenzenes, and many other liquids. A linear relationship between the cosine of the contact angle ($\cos\theta$) and the surface tension of the liquid (γ_{LV}) for the n-alkane solvents was obtained (Fig. 7.2). Similar plots were also obtained for di(n-alkyl)ethers and

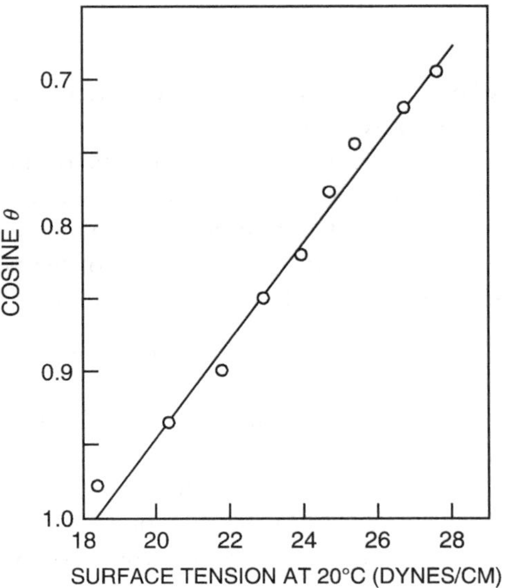

Fig. 7.2 Plot of cosθ against surface tension for a series of *n*-alkanes on PTFE (reproduced with permission from [8], copyright 1950 Elsevier)

n-alkylbenzenes [9]. When the plot in Fig. 7.2 is extended to a much larger solvent set, including other halocarbons, ketones, esters, amides, and water, the plot is no longer linear (see Fig. 3.15 in Chap. 3). Zisman [9] attributed the linear portion of the relationship to the van der Waals force interaction between *n*-alkanes and the PTFE surface. The deviation from linearity for high surface tension liquids was attributed to stronger intermolecular interactions, such as dipolar interactions and H-bonding, in these solvents. The intercept at cosθ = 1.0 is at ~18 mN/m and is defined as critical surface tension (γ_C) of the solid. Liquids with surface tensions lower than γ_C will fully wet the solid surface, while liquids with higher surface tensions than γ_C will form a finite contact angle on the solid surface. In other words, the critical surface tension is the highest surface tension which the liquid fully wets the solid surface. This critical surface tension is found to be empirically close to the solid surface tension. The plot in Fig. 7.2 is now known as the Zisman plot in literature and has become one of the methods to estimate the surface tension of solid surface.

Surface wettability depends strongly on molecular interactions between liquid and the solid surface. In 1953, Ellison, Fox, and Zisman [10] reported a wetting study of three adsorbed monolayers with different end groups with *n*-alkanes, along with a PTFE film. The end groups from the adsorbed monolayers are CH_3, CHF_2, and CF_3. The main functional group on the PTFE film is CF_2. The Zisman plot for the study is depicted in Fig. 7.3. From the intercepts at cosθ = 1.0, these authors found that the critical surface tensions of the surface constituents decrease in the following order: $CH_3 > CF_2 > CHF_2 > CF_3$. From the intercept at cosθ = 1, the critical surface tension for the CF_3 head group was determined to be ~6 mN/m. This is very similar to the surface free energy value of *n*-perfluoroeicosane ($C_{20}F_{42}$), which is

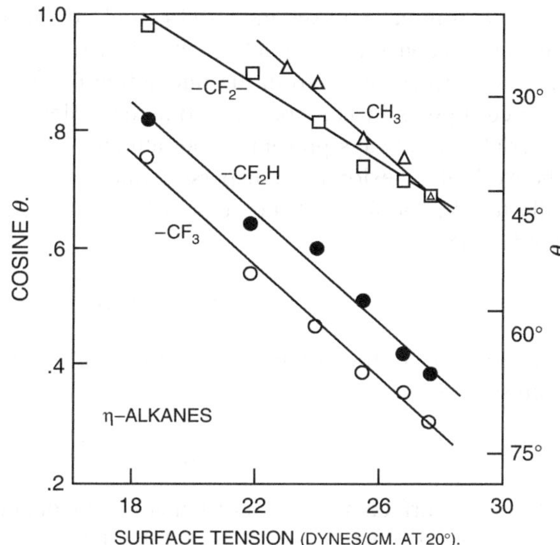

Fig. 7.3 Plot of wettability data for adsorbed monolayers with CH₃, CF₂H, and CH₃ end group and PTFE film with *n*-alkanes (reproduced with permission from [10], copyright 1953 The American Chemical Society)

6.7 mN/m, determined by Nishino et al. using the dynamic contact angles of water and diiodomethane [11]. The results demonstrate that the Zisman plot can be used to determine the surface tension of different chemical constituents on the solid surface. Indeed, the critical surface tensions of different chemical constituents from surfaces of fluorocarbon, hydrocarbon, chlorocarbon, and nitrated hydrocarbon polymers have been determined [12].

It is important to point out that the Zisman critical surface tension analysis is only limited to low surface energy surfaces and alkane solvents, where van de Waals force is the prime interaction. Many exceptions were found with polar and H-bonding solvents. Despite this limitation, the Zisman analysis remains as one of the useful tools to estimate the surface tension of low surface energy material to date [13].

7.2.2 Surface Tension Component Methods

Fowkes method. Fowkes [14–16] was the pioneer proposing partition of surface tension. He assumed that the surface tension of a solid or liquid is a sum of independent components, which addresses specific molecular interaction individually. For example, γ_{SV} can have a number of contributors and is given by

$$\gamma_{SV} = \gamma_{SV}^{d} + \gamma_{SV}^{p} + \gamma_{SV}^{h} + \gamma_{SV}^{i} + \gamma_{SV}^{ab} \tag{7.5}$$

where γ_{SV}^{d}, γ_{SV}^{p}, γ_{SV}^{h}, γ_{SV}^{i}, γ_{SV}^{ab} represent contributions from the dispersion, dipole–dipole interaction, hydrogen bonding, induced dipole–dipole interaction, and acid–base components, for example.

In the original Fowkes model [14], only dispersion component of the surface tension was considered, which is caused by London dispersion force. The London dispersion forces arise from the interaction of fluctuating electronic dipoles with induced dipoles in neighboring atoms or molecules [15]. It exists in all type of materials and always presents as an attractive force at the liquid–solid interface. The work of adhesion from dispersion interaction has been proved thermodynamically to take the form of the geometric mean according to the Berthelot mixing rule [17, 18].

$$W_{SL}^{d} = \sqrt{W_{SV}^{d} \cdot W_{LV}^{d}} \tag{7.6}$$

Therefore, according to Eq. (7.3), the liquid–solid interfacial tensions can be expressed as

$$\gamma_{SL} = \gamma_{SV} + \gamma_{LV} - 2\sqrt{\gamma_{SV}^{d} \cdot \gamma_{LV}^{d}} \tag{7.7}$$

If the solid surface has only a dispersion component, $\gamma_{SV} = \gamma_{SV}^{d}$. By combining the Young–Dupre equation (Eq. 7.4) and the definition of work of adhesion (Eq. 7.2), γ_{SV} can be expressed as

$$\gamma_{SV} = \gamma_{SV}^{d} = \frac{\gamma_{LV}^{2}}{4\gamma_{LV}^{d}} \cdot \left(1 + \cos\theta\right)^{2} \tag{7.8}$$

From Eq. (7.8), the solid surface tension (γ_{SV}) can be calculated by measuring the contact angle θ with a liquid whose γ_{LV} value is known.

Owens–Wendt–Rabel–Kaelble (OWRK) method. Owens and Wendt [19] modified the Fowkes model by assuming that solid surface tension and liquid surface tensions are composed of two components, namely, a dispersion component and a hydrogen-bonding component. The nondispersive interaction was included into the hydrogen-bonding component. Nearly at the same time, Rabel [20] and Kaelble [21] also published a very similar equation by partitioning the solid surface tension into dispersion and polar components. Subsequent researchers called this as the OWRK method, and γ_{SV} and γ_{LV} can be expressed as

$$\gamma_{SV} = \gamma_{SV}^{d} + \gamma_{SV}^{p}$$

$$\gamma_{LV} = \gamma_{LV}^{d} + \gamma_{LV}^{p} \tag{7.9}$$

Similar to Eq. (7.7), the interfacial liquid–solid surface tension can be derived by assuming that the polar component has a geometric mean form, although dipole–dipole interactions follow the rule of geometric mean [15] and hydrogen-bonding interactions are probably not [15, 16, 22].

$$\gamma_{SL} = \gamma_{SV} + \gamma_{LV} - 2\sqrt{\gamma_{SV}^{d} \cdot \gamma_{LV}^{d}} - 2\sqrt{\gamma_{SV}^{p} \cdot \gamma_{LV}^{p}} \tag{7.10}$$

By combining with the Young–Dupre equation (Eq. 7.4), we have

$$\gamma_{LV}\left(1+\cos\theta\right)=2\sqrt{\gamma_{SV}^{d}\cdot\gamma_{LV}^{d}}+2\sqrt{\gamma_{SV}^{p}\cdot\gamma_{LV}^{p}} \tag{7.11}$$

There are two unknowns, γ_{SV}^{d} and γ_{SV}^{p}, in Eq. (7.11). They can be calculated by determining the contact angles with two different test liquids with known γ_{LV}^{d} and γ_{LV}^{p} values. γ_{SV} will be just the sum of γ_{SV}^{d} and γ_{SV}^{p}. Since two test liquids are required, this method is also called the two-liquid method.

Extended Fowkes method. Kitazaki et al. [23] further refined the OWRK model by splitting the dipole–dipole interactions into polar and H-bonding interactions. Accordingly, both γ_{SV} and γ_{LV} comprise three components: d for dispersion, p for polar, and h for H-bonding (Eq. 7.13), and they all take the form of geometrical mean (Eqs. 7.14 and 7.15).

$$\gamma_{SV}=\gamma_{SV}^{d}+\gamma_{SV}^{p}+\gamma_{SV}^{h}$$

$$\gamma_{LV}=\gamma_{LV}^{d}+\gamma_{LV}^{p}+\gamma_{LV}^{h} \tag{7.13}$$

$$\gamma_{SL}=\gamma_{SV}+\gamma_{LV}-2\sqrt{\gamma_{SV}^{d}\cdot\gamma_{LV}^{d}}-2\sqrt{\gamma_{SV}^{p}\cdot\gamma_{LV}^{p}}-2\sqrt{\gamma_{SV}^{h}\cdot\gamma_{LV}^{h}} \tag{7.14}$$

$$\gamma_{LV}\left(1+\cos\theta\right)=2\sqrt{\gamma_{SV}^{d}\cdot\gamma_{LV}^{d}}+2\sqrt{\gamma_{SV}^{p}\cdot\gamma_{LV}^{p}}+2\sqrt{\gamma_{SV}^{h}\cdot\gamma_{LV}^{h}} \tag{7.15}$$

There are three unknowns in the solid surface tension, γ_{SV}^{d}, γ_{SV}^{p}, and γ_{SV}^{h} in Eq. (7.15). To calculate γ_{SV}, one can use three test liquids with known γ_{LV}^{d}, γ_{LV}^{p}, and γ_{LV}^{h} values. The component solid surface tensions can be determined from the three contact angles, and γ_{SV} is simply the sum of the three components. This method is also known as the three-liquid method.

van Oss, Chaudhury, and Good (vOCG) model. van Oss et al. [24–26] proposed a slightly different approach to partition the surface tensions in order to address hydrogen-bonding interactions. In the vOCG model, the surface tensions compose of (a) an apolar component of the interfacial tension γ^{LW} (Lifshitz–van der Waals interactions, including dispersion, dipole–dipole interaction, and induced dipole–dipole interactions) and (b) a short range surface tension γ^{AB} including hydrogen-bonding interactions. The solid and liquid surface tensions are given by

$$\gamma_{SV}=\gamma_{SV}^{LW}+\gamma_{SV}^{AB}$$

$$\gamma_{LV}=\gamma_{LV}^{LW}+\gamma_{LV}^{AB} \tag{7.16}$$

where $\gamma_{SV}^{AB}=2\sqrt{\gamma_{SV}^{+}\cdot\gamma_{SV}^{-}}$ and $\gamma_{LV}^{AB}=2\sqrt{\gamma_{LV}^{+}\cdot\gamma_{LV}^{-}}$. γ_{SV}^{+} and γ_{SV}^{-} γ_{LV}^{+} and γ_{LV}^{-} are the electron acceptor (Lewis acid) and electron donor (Lewis base) for the solid and liquid, respectively.

The work of adhesion W_{SL}^{AB} and the interfacial surface tensions γ_{SL}^{AB} and γ_{SL} can be derived from molecular orbital theory [27, 28]:

$$W_{SL}^{AB}=2\sqrt{\gamma_{SV}^{+}\cdot\gamma_{LV}^{-}}+2\sqrt{\gamma_{SV}^{-}\cdot\gamma_{LV}^{+}} \tag{7.17}$$

$$\gamma_{SL}^{AB} = \gamma_{SV}^{AB} + \gamma_{LV}^{AB} - 2\sqrt{\gamma_{SV}^{+} \cdot \gamma_{LV}^{-}} - 2\sqrt{\gamma_{SV}^{-} \cdot \gamma_{LV}^{+}} \tag{7.18}$$

$$\gamma_{SL} = \gamma_{SV} + \gamma_{LV} - 2\sqrt{\gamma_{SV}^{LW} \cdot \gamma_{LV}^{LW}} - 2\sqrt{\gamma_{SV}^{+} \cdot \gamma_{LV}^{-}} - 2\sqrt{\gamma_{SV}^{-} \cdot \gamma_{LV}^{+}} \tag{7.19}$$

From the Young–Dupre equation in Eq. (7.4), the relationship between contact angle and the surface tension components can be written as

$$\gamma_{LV}\left(1 + \cos\theta\right) = 2\sqrt{\gamma_{SV}^{LW} \cdot \gamma_{LV}^{LW}} + 2\sqrt{\gamma_{SV}^{+} \cdot \gamma_{LV}^{-}} + 2\sqrt{\gamma_{SV}^{-} \cdot \gamma_{LV}^{+}} \tag{7.20}$$

There are three unknowns in the solid surface tension, γ_{SV}^{LW}, γ_{SV}^{+}, and γ_{SV}^{-}. Experimentally, they can be calculated by measuring the contact angles with three liquids with known γ_{LV}^{LW}, γ_{LV}^{+}, and γ_{LV}^{-} values.

In addition to the methods discussed above, there are a few other solid surface tension determination methods, such as the Wu method [29, 30] and the Schultz methods [31, 32], which also fall into the category of partitioning surface tensions into independent components. Wu used the harmonic means to describe the interfacial surface tension instead of the geometric mean, based on a few slightly different assumptions to derive the equations for Wu's model. The Schultz methods can be considered as a special case of the extended Fowkes method. The contact angle of a polar liquid (usually water) on the solid is conducted in another nonpolar liquid medium (e.g., pure hydrocarbon compounds), or the contact angle of a nonpolar liquid on the solid is measured in another polar liquid medium.

7.2.3 Equation of State

The method of equation of state is totally different from all the surface tension component methods described in Sect. 7.2.2. The equation of state method assumes that the interfacial surface tension γ_{SL} depends on the surface tension of the liquid γ_{LV} and solid γ_{SV} only, i.e., $\gamma_{SL} = f\left(\gamma_{SV}, \gamma_{LV}\right)$. The method was mainly developed in Neumann's laboratory [33–37]. Neumann et al. formulated three versions of the equation of state. The first version was based on Zisman's data comprising eight solid surfaces with low surface tensions [33, 34]. According to Girifalco and Good [38], the liquid–solid interfacial surface tension between dissimilar molecules can be formulated as

$$\gamma_{SL} = \gamma_{SV} + \gamma_{LV} - 2\varphi_{SL} \cdot \sqrt{\gamma_{SV} \cdot \gamma_{LV}} \tag{7.21}$$

φ_{SL} is a characteristic constant of the system and is equaled to one when the interactions are from similar types of molecules and not much difference between γ_{SV} and γ_{LV}, which is the Berthelot's combining rule. However, in most situations φ_{SL} is an unknown when measuring solid surface tension. Neumann et al. [33, 34] performed a curve fitting to obtain the relationship of φ_{SL} and interfacial tension γ_{SL} by assuming that solid surface tension equals to the critical surface tension determined from the Zisman plot (Eq. 7.22). Eventually the explicit expression of γ_{SL} is derived by combining Eqs. (7.21) and (7.22).

$$\varphi_{SL} = -0.00775\gamma_{SL} + 1 \tag{7.22}$$

$$\gamma_{SL} = \frac{\left(\sqrt{\gamma_{LV}} - \sqrt{\gamma_{SV}}\right)^2}{1 - 0.015\sqrt{\gamma_{SV} \cdot \gamma_{LV}}} \tag{7.23}$$

Then this relationship can be used to determine the solid surface tension (γ_{SV}) using only one liquid with known surface tension.

There are two other equations of state formulated from larger data set of surfaces and testing liquids [35–37].

$$W_{SL} = 2 \cdot \sqrt{\gamma_{SV} \cdot \gamma_{LV}} \cdot e^{-\beta(\gamma_{LV} - \gamma_{SV})^2} \tag{7.24}$$

$$\gamma_{SL} = \gamma_{SV} + \gamma_{LV} - 2 \cdot \sqrt{\gamma_{SV} \cdot \gamma_{LV}} \cdot e^{-\beta(\gamma_{LV} - \gamma_{SV})^2} \tag{7.25}$$

where $\beta = 0.0001247\,(m^2/mJ)^2$.

$$W_{SL} = 2 \cdot \sqrt{\gamma_{SV} \cdot \gamma_{LV}} \cdot \left(1 - \beta_1 \cdot \left(\gamma_{LV} - \gamma_{SV}\right)^2\right) \tag{7.26}$$

$$\gamma_{SL} = \gamma_{SV} + \gamma_{LV} - 2 \cdot \sqrt{\gamma_{SV} \cdot \gamma_{LV}} \cdot \left(1 - \beta_1 \cdot \left(\gamma_{LV} - \gamma_{SV}\right)^2\right) \tag{7.27}$$

where $\beta_1 = 0.0001057\,(m^2/mJ)^2$.

The coefficients β and β_1 are determined experimentally by measuring the contact angles on 15 solid surfaces with a series of testing liquids. Kwok and Neumann [36, 37] also suggested to adjust the coefficients and the solid surface tension simultaneously to get more precise calculation of the solid surface tension. They further demonstrated that despite the different formulations and coefficients, the three equations of states have yielded the same γ_{SV} values, based on various set of experimental contact angles [36, 37].

7.2.4 Assessment of the Different Methods

In summary, there are three basic approaches to use contact angle data to determine the surface tensions of solid surfaces. These approaches are the Zisman method, the surface tension component methods, and the equation of state. Within these three approaches, there are many variants. It is reasonable to wonder the merit, accuracy, and limitation of some of the methods. The Zisman method is an empirical approach based on the correlation between the cosines of the contact angles on a solid surface versus the surface tensions of the test liquids. With alkanes, linear plots are usually obtained, and the critical solid surface tension (γ_C) is determined by extrapolating the plot to $\cos\theta = 1.0$. Zisman never claimed γ_C is γ_{SV}, although some in the literature assumed so. The analysis is limited to alkanes as the plots tend to depart from linear due to polar and H-bonding interactions. Zisman et al. also observed many exceptions during their studies with higher energy surfaces owing to specific interactions. In a sense, the Zisman approach is limited to low surface energy surfaces where the van de Waals interaction is dominant [9].

The differences between the surface tension component methods and the equation of state method are very distinct, and there have been significant debates in the literature between these two approaches [36, 37, 39–45]. The equation of state method is an empirical fitting method by assuming that the liquid–solid interfacial surface tension is a function of the solid and liquid surface tension only. The thermodynamic argument of the approach has been heavily criticized first by Morrison [42], who conducted a thermodynamic analysis and concluded that the equation of state is thermodynamically erroneous. Siboni and coworkers [45] later questioned the thermodynamic justification of the theory, particularly the calculation of the degree of freedom for a two-component three-phase system, as well as the approximation used in deriving the equation. They concluded that both of the theory and the approximation are highly questionable, if not wrong. In fact, Good felt so strongly about the incorrectness of the approach that he withdrew his support of the work he coauthored earlier [18, 34].

From the molecular interaction standpoint, the equation of state did not consider any polar, hydrogen bond, or acid–base interactions in formulating φ_{SL}. Therefore, it may be inadequate to illustrate the interfacial tensions between the solid and liquid phases [18]. Strong evidences have been reported for the existence of molecular interactions at the liquid–solid interface. For example, Fowkes et al. [40] were able to observe interfacial van der Waals and Lewis acid–base interactions directly as chemical shifts in infrared and NMR spectra. These are the direct evidences that γ_{SL} should be a function of some molecular interactions, not just γ_{SV} and γ_{LV} as shown in Eq. (7.21). Inspired by the interactions between two immiscible polar liquids [38], theoretical attempts were made to predict the value of φ_{SL} from the molecular properties, e.g., dipole moment, polarizability, and ionization energy of the solid and liquid phases [18, 41]. Although promising results were obtained, notable exceptions to the correlation, e.g., aromatic compounds, could not be resolved except by involving interactions with hydrogen bonds [41].

The surface tension component method assumes that surface tension can be partitioned into different components, which address different intermolecular interactions individually. The overall surface tension will be the sum of all the components according to the linear free energy relationship. In the original Fowkes method [14], only the dispersion interaction is considered. The component method has been subsequently extended to include polar component and later divided the polar component into dipolar interaction and H-bonding interaction. The vOCG model appeared as a refined version of the surface tension component methodology. It assumes the existence of both additive and nonadditive components. The Lifshitz–van der Waals component (γ^{LW}) is additive, and the electron donor and acceptor components (γ^- and γ^+) are nonadditive. The solid surface tension (γ_{SV}) can be calculated by using three liquids with known γ^{LW}, γ^- and γ^+ values. Since this is a semiempirical approach, the calculated value may vary a little depending on the choice of liquid set [44]. In general, the vOCG method considers all type of molecular interactions and is probably suitable for general use. If there is certainty that the solid surface has dispersion component only, the Fowkes model and the Zisman plot can be used as well.

7.3 Fundamental Issues and Outlook

7.3.1 Solid Surface Energy Versus Surface Tension

Due to simplicity, contact angle measurement has been widely used to determine surface energy of solid. It is important to point out that most people are interested in the surface energy of the solid not surface tension. The assumption is that the higher the surface energy, the stickier the surface and vice versa. Johnson [7] noted in 1959 that the terms surface energy, surface free energy, and surface tension had been used interchangeably and as desired by authors. This practice unfortunately has not been changed today. In any event, surface energy and surface tension are two distinct quantities in solid [5–7], and what's determined from contact angle measurement is surface tension. The cause of this confusion can be traced back to Dupre [3], who hypothesized the reversibility of the work of adhesion between two dissimilar materials. If these two materials are immiscible liquids such as that shown in Fig. 7.1b, the free energy change per unit area at the interface would be equaled to the surface tension of the newly formed interface minus the surface tensions of the two precursor liquids (Eq. 7.3). Unlike liquid, molecules at the surface or interface of a solid have little mobility during interfacial interactions. The extension of the work-of-adhesion assumption to a one-solid, one-liquid system may not be justifiable. In fact, Girifalco and Good [38] had pointed this out nearly six decades ago. The solid surface tension (γ_{SV}) value calculated from contact angle is hence questionable.

7.3.2 Which Contact Angle One Should Use?

As pointed out in Chap. 3, there are at least four measurable contact angles for a given liquid–solid system. They are the static (or Young's) angle θ, the advancing and receding angle θ_A and θ_R, and the equilibrium angle θ_{eq}. The first three angles are from the metastable wetting states. The latter is from a thermodynamically stable state, which is populated through appropriate vibration of the Young's sessile droplet. Survey of the solid surface tension literature suggests that most of the solid surface tensions were determined from θ, which is from a metastable state. Should one recalculate all the surface tension numbers using θ_{eq} as suggested by Marmur et al. [46] and Della Volpe et al. [47]? This really challenges the legitimacy of all solid surface tension numbers in the literature.

7.3.3　Path Forward

The motivation of determining solid surface energy or more precisely solid surface tension is to use the number to predict wettability and adhesion and sometimes infer molecular interactions. While contact angle is a very simple measurement, as briefly overviewed in Sect. 7.2, the methodologies employed to calculate the surface tension are actually not straightforward. They are mostly semiempirical. There are assumptions over assumptions and some of which have yet to be validated! The fundamental question one may ask is: is surface tension the correct quantity one needs to predict wettability and adhesion? We actually got the answer from Fowkes, who pioneered the surface tension component approach. He concluded in one of his papers that surface energy is actually not of any importance for surface and interfacial studies [40]. Using the microbalance technique, Samuel, Zhao, and Law [48] showed that surface wettability can be predicted from the advancing contact angle θ_A, the larger the θ_A, the lower the wettability. In the same study, they also showed that surface adhesion is correlating to θ_R, the smaller the θ_R, the stronger the surface adhesion. Since the whole objective of determining surface tension (energy) is to be able to predict surface wettability and adhesion interaction, we recommend everyone to simply measure θ_A and θ_R instead.

References

1. Cuenot S, Fretigny C, Demoustier-Champagne S, Nysten B (2004) Surface tension effect on the mechanical properties of nanomaterials measured by atomic force microscopy. Phys Rev B Condens Matter 69:165410
2. Young T (1805) An essay on the cohesion of fluids. Philos Trans R Soc Lond 95:65–87
3. Dupre A (1869) Theorie Mechanique de la Chaleur. Gauthier-Villars, Paris, p 369
4. Schrader ME (1995) Young-dupre revisited. Langmuir 11:3585–3589
5. Hui CY, Jagota A (2013) Surface tension, surface energy, and chemical potential due to their difference. Langmuir 29:11310–11316
6. Gray VR (1965) Surface aspects of wetting and adhesion. Chem Ind 23:969–977
7. Johnson RE (1959) Conflicts between Gibbsian thermodynamics and recent treatments of interfacial energies in solid–liquid-vapor systems. J Phys Chem 63:1655–1658
8. Fox HW, Zisman WA (1950) The spreading of liquids on low energy surfaces. I. Polytetrafluoroethylene. J Colloid Sci 5:514–531
9. Zisman WA (1964) Relation of the equilibrium contact angle to liquid and solid constitution. In: Fowkes F (ed) Contact angle, wettability, and adhesion, advances in chemistry. American Chemical Society, Washington, DC, pp 1–51
10. Ellison AH, Fox HW, Zisman AW (1953) Wetting of fluorinated solids by H-bonding liquids. J Phys Chem 57:622–627
11. Nishino T, Meguro M, Nakamae K, Matsushita M, Ueda Y (1999) The lowest surface free energy based on -CF3 alignment. Langmuir 15:4321–4323
12. Shafrin EG, Zisman WA (1960) Constitutive relations in the wetting of low energy surfaces and the theory of the retraction method of preparing monolayer. J Phys Chem 64:519–524
13. Chhatre SS, Guardado JO, Moore BM, Haddad TS, Mabry JM, McKinley GH, Cohen RE (2010) Fluoroalkylated silicon-containing surfaces—estimation of solid-surface energy. ACS Appl Mater Interfaces 2:3544–3554

14. Fowkes FM (1962) Determination of interfacial tensions, contact angles, and dispersion forces in surfaces by assuming additivity of intermolecular interactions in surface. J Phys Chem 66:382

15. Fowkes FM (1964) Attractive forces at interfaces. Ind Eng Chem 56:40–52

16. Fowkes FM (1972) Donor-acceptor interactions at interfaces. J Adhes 4:155–159

17. Berthelot D (1898) Sur le mélange des gaz. C R Hebd Seances Acad Sci 126:1857–1858

18. Good RJ (1992) Contact angle, wetting, and adhesion: a critical review. J Adhes Sci Technol 6:1269–1302

19. Owens DK, Wendt RC (1969) Estimation of the surface free energy of polymers. J Appl Polym Sci 13:1741–1747

20. Rabel W (1971) Einige Aspekte der Benetzungstheorie und ihre Anwendung auf die Untersuchung und Veränderung der Oberflächeneigenschaften von Polymeren. Farbe und Lack 77:997–1005

21. Kaelble DH (1970) Dispersion-polar surface tension properties of organic solids. J Adhes 2:66–81

22. Janczuk B, Bialopiotrowicz T (1989) Surface free-energy components of liquids and low energy solids and contact angles. J Colloid Interface Sci 127:189–204

23. Kitazaki Y, Hata TJ (1972) Surface-chemical criteria for optimum adhesion. J Adhes 4:123–132

24. Van Oss CJ, Good RJ, Chaudhury MK (1986) The role of van der Waals forces and hydrogen bonds in "hydrophobic interactions" between biopolymers and low energy surfaces. J Colloid Interface Sci 111:378–390

25. Van Oss CJ, Ju L, Chaudhury MK, Good RJ (1988) Estimation of the polar parameters of the surface tension of liquids by contact angle measurements on gels. J Colloid Interface Sci 128:313–319

26. van Oss CJ (2006) Interfacial forces in aqueous media. Taylor & Francis, New York

27. Kollman P (1977) A general analysis of noncovalent intermolecular interactions. J Am Chem Soc 99:4875–4894

28. Hobza P, Zahradnik R (1980) Weak intermolecular interactions in chemistry and biology. Elsevier, New York

29. Wu S (1971) Calculation of interfacial tension in polymer system. J Polym Sci C 34:19–30

30. Wu S (1973) Polar and nonpolar interactions in adhesion. J Adhes 5:39–55

31. Schultz J, Tsutsumi K, Donnet JB (1977) Surface properties of high-energy solids I. Determination of the dispersive component of the surface free energy of mica and its energy of adhesion to water and n-alkanes. J Colloid Interface Sci 59:272–277

32. Schultz J, Tsutsumi K, Donnet JB (1977) Surface properties of high-energy solids determination of the nondispersive component of the surface free energy of mica and its energy of adhesion to polar liquids. J Colloid Interface Sci 59:278–282

33. Sell PJ, Neumann AW (1966) The surface tension of solids. Angew Chem Int Ed 5:299–307

34. Neumann AW, Good RJ, Hope CJ, Sejpal M (1974) An equation-of-state approach to determine surface tensions of low-energy solids from contact angles. J Colloid Interface Sci 49:291–304

35. Li D, Neumann AW (1990) A reformulation of the equation of state for interfacial tensions. J Colloid Interface Sci 137:304–307

36. Kwok DY, Neumann AW (1999) Contact angle measurement and contact angle interpretation. Adv Colloid Interface Sci 81:167–249

37. Kwok DY, Neumann AW (2000) Contact angle interpretation in terms of solid surface tension. Colloids Surf A Physicochem Eng Asp 161:31–48

38. Girifalco LA, Good RJ (1957) A theory for the estimation of surface and interfacial energies. I. Derivation and application to interfacial tension. J Phys Chem 61:904–909

39. Van Oss CJ, Good RJ, Chaudhury MK (1988) Additive and nonadditive surface tension components and the interpretation of contact angles. Langmuir 4:884–891

40. Fowkes FM, Riddle FL, Pastore WE, Weber AA (1990) Interfacial interactions between self-associated polar liquids and squalane used to test equations for solid–liquid interfacial interactions. Colloids Surf 43:367–387

41. Good RJ, Elbing E (1971) Generalization of theory for estimation of interfacial energies. Chem Phys Interfaces 2:72–96
42. Morrison I (1991) Does the phase rule for capillary systems really justify an equation of state for interfacial tensions? Langmuir 7:1833–1836
43. Zenkiewicz M (2007) Comparative study on the surface free energy of a solid calculated by different methods. Polym Test 26:14–19
44. Della Volpe C, Maniglio D, Brugnara M, Siboni S, Morra M (2004) The solid surface free energy calculation I. In defense of the multicomponent approach. J Colloid Interface Sci 271:434–453
45. Siboni S, Della Volpe C, Maniglio D, Brugnara M (2004) The solid surface free energy calculation II. The limits of the Zisman and of the "equation-of-state" approaches. J Colloid Interface Sci 271:454–472
46. Cwikel D, Zhao Q, Liu C, Su X, Marmur A (2010) Comparing contact angle measurements and surface tension assessments of solid surfaces. Langmuir 26:15289–15294
47. Della Volpe C, Maniglio D, Morra M, Siboni S (2002) The determination of a "stable equilibrium" contact angle on heterogeneous and rough surfaces. Colloids Surf A Physicochem Eng Asp 206:47–67
48. Samuel B, Zhao H, Law KY (2011) Study of wetting and adhesion interactions between water and various polymer and superhydrophobic surfaces. J Phys Chem C 115:14852–14861

Chapter 8
Summary and Final Remarks

Abstract Many believe that the concept of wetting begins with the Young's equation. Although the Young's equation is very simple, it has been a source of arguments over the last two centuries because the equation comprises four quantities of which two of them cannot be measured reliably. Moreover, researchers did express frustration in their inability to measure the Young's angle consistently, at least a century ago. This chapter provides a brief overview of the history and the source of some of the misconceptions. Fundamental concepts that have been clarified in recent years, including (1) the recognition of the fact that it is the contact line, not the contact area, that determines the contact angle; (2) advancing and receding contact angles are the most important contact angles, and they measure wettability and adhesion respectively, and surface stickiness can be predicted from the sliding angle; and (3) hydrophilicity and hydrophobicity should be defined by the receding contact angle, not the static contact angle. In answering Good's calling for standardization of measurement protocols for contact angle measurements, a set of guidelines for determining static contact angle, advancing/receding contact angle, and sliding angle are provided. We hope that these guidelines will benefit the community in the near term and serve as a springboard for the development of standardized procedures by the "authority" or leaders in this field in the near future.

Keywords Young's equation • Misconceptions • Mechanical equilibrium • Young's angle • Advancing contact angle • Receding contact angle • Contact angle hysteresis • Ideal surface • Real surface • Contact line • Contact area • Surface characterization • Measurement protocols • Guidelines and best practices

8.1 Misconceptions in the Young's Equation

Most researchers would agree that surface science started with Thomas Young when he published his legendary article in 1805 entitled "An Essay on the Cohesion of Fluids" [1]. Young did not write any equation in his paper, but he descriptively stated that an angle is formed when liquid wets a solid surface. He wrote: *We may therefore inquire into the conditions of equilibrium of the forces acting on the angular particles, one in the direction of the surface of the fluid only, a second in that of the common surface of the solid and fluid, and the third in that of the exposed surface of the solid.* This is essentially the Young's equation we know to date.

© Springer International Publishing Switzerland 2016

K.-Y. Law, H. Zhao, *Surface Wetting*, DOI 10.1007/978-3-319-25214-8_8

Young clearly stated that the angle of contact is a result of the mechanical equilibrium among the three surface tensions (γ_{LV}, γ_{SL}, and γ_{SV}) at the three-phase contact line. However, it is somewhat a mystery that there are still misquotations in recent years, either by implying or directly stating, that the contact angle is from a thermodynamically equilibrium wetting state. Such a quotation is not correct.

The Young's equation is deceptively simple, but it is problematic multidimensionally. First, two out of four quantities in the equation, γ_{SV} and γ_{SL}, cannot be measured reliably. According to Zisman [2], researchers in the nineteenth century, notably Dupre [3] and Gibbs [4], had resorted to thermodynamics to solve the Young's equation. Due to the lack of available documents in this period, we are unable to map out the evolution of the history of surface research in that era. It is however likely that our bias toward thermodynamics may be a result of our basic science training in school. As discussed in Chaps. 3 and 4, although the process of wetting is driven by thermodynamics, the way the wetting liquid reaches its final static state is controlled by kinetics. In other words, the Young's angle is from a metastable wetting state. Most static contact angles on flat surfaces and rough surfaces, including the Wenzel angle and the Cassie-Baxter angle, are all from the metastable wetting states. This has been one of the reasons for the continuous argument as researchers had a hard time to measure the Young's angle on a consistent basis.

The second issue around the Young's equation is the Young's angle itself. Researcher as early as 1890 [5] had reported the existence of multiple angles between the advancing and receding angles, which could be reliably measured. The Young's angle is just one of the angles between advancing and receding and was known to be a metastable state for some time [6–9]. Bartell and Wooley even suggested that the receding angle may be the most important angle among the three angles they measured [9]. In his 1959 paper, Johnson pointed out that researchers during that time had serious reservation about the validity of the Young's equation [10]. Experimental evidence in the last couple of decades, which is summarized in Chap. 3, shows that there indeed exist multiple metastable wetting states between the advancing and receding angles. They can be populated by carefully exciting the sessile droplet with vibration noise (Fig. 3.11). A key message from these studies is that, all the metastable wetting states can be driven to a thermodynamically most stable wetting states with optimal vibration excitation [11–14].

The third issue derived from the Young's equation is contact angle hysteresis (CAH), the difference between advancing and receding contact angle. Some early researchers thought CAH originated from microscopic heterogeneity on the surface, particularly roughness. Some thought the Young's equation would only be applicable to homogeneous, smooth, and rigid surfaces, and the term ideal surface was created. To date, we know this is not true. Both roughness [15–17] and molecular interactions between liquid and the solid surface [18, 19] are contributors to CAH, and an update was provided in Chap. 5 in this book. Undoubtedly, more work is needed to thoroughly understand the mechanism that leads to hysteresis and the role it may play in surface science in general.

8.2   Concepts that Are Turning the Corners

The pioneering thermodynamic analyses of the wetting process by Dupre [3] and Gibbs [4], coupled with our basic science training in school, seem to create a strong bias in solving scientific problem based on thermodynamic principles. The successful work by Wenzel [20] to explain the enhancement of wettability for rough hydrophilic surface and increase of resistance to wet for hydrophobic surface, coupled with the work by Cassie and Baxter [21] on porous surfaces, further solidified our bias. Although Pease [6] and Johnson [10], in the same era, had voiced the importance of the contact line, rather than the wetted area dictates the contact angle, their messages had been largely ignored for decades. On the other hand, mounting results in the last two decades, both experimental and theoretical, show convincingly that *contact angle is a one-dimensional, not two-dimensional problem. Contact angle is dictated by the locality of the three-phase contact line, not the area underneath the liquid droplet.* While disagreements between the calculated Wenzel and Cassie angles with the experimentally observed angles are often found, the discrepancy can be resolved when the classic roughness factors are modified with correction factors that include the geometry of the rough contact line [22–24]. It is important to point out that agreements between the calculated Wenzel and Cassie-Baxter angles with the observed angles are possible. This can occur when the roughness is microscopic and homogeneous such that the roughness factor calculated from the wetted area is comparable to that at the three-phase contact line [25].

Many leading researchers had advocated the importance of θ_A and θ_R in surface study, the fundamental reason was never offered clearly or convincingly. It was either because of the CAH or the surface is not *real*. Using the microbalance technique, Samuel et al. [26] reported the measure of the wetting force and the adhesion force between water and 20 surfaces of varying wettability. Good correlations were found between the wetting force and θ_A and the adhesion force and θ_R. This work establishes a clear connection between contact angles and liquid–solid interactions. The study led to the recommendation that θ_A is a measure of surface wettability, the smaller the θ_A, the higher the wettability. Similarly θ_R can be used to predict surface adhesion, the smaller the θ_R, the stronger the surface adhesion. Sliding angle α determines the mobility of the drop or the stickiness of the surface. Evidence is provided that CAH is neither correlating to wettability nor adhesion and stickiness (Figs. 5.2, 5.7 and 5.14).

Hydrophilicity and hydrophobicity are the most recognized terms in surface science. Using them as adjectives and comparing affinity is all right. However, the science community has generally accepted that a surface is hydrophilic when its water contact angle is <90°, and it is hydrophobic when the water contact angle is >90°. There is little technical evidence to support this definition. Worse yet, the 90° cutoff has been extended to define the philicity/phobicity boundaries of other liquids. Bad definition does have unwanted consequences. For example, after defining $\theta > 90°$ as hydrophobic, researchers further come to the consensus arbitrarily that superhydrophobicity is defined for surfaces having a contact angle >150° [27].

The inaccuracy of these definitions has led to confusing terminologies, such as sticky superhydrophobicity, oleophobic wettability, and so on. A surface is defined as superhydrophobic when it exhibits high water repellency. Can a sticky surface exhibit high repellency at the same time? Sticky superhydrophobicity seems novel, but the two components in the terminology are actually contradicting to each other. After a very careful analysis of a series of water contact angle data along with the correlations with the wetting force and adhesion force, Law [28, 29] showed that hydrophilicity and hydrophobicity should be defined by the receding contact angle. The cutoff is at water θ_R 90°. More importantly, the fundamental reason why the surface changes its character, from hydrophilic to hydrophobic, is a result of the stronger cohesion force within the liquid droplet versus the wetting force due to wetting. For hydrophobic surface, the water drop prefers to be in the droplet state than wetting a hydrophobic surface due to the weak wetting interaction. Since Samuel et al. also showed that there is negligible attraction between water and surfaces with $\theta_A \geq 145°$, a surface can further be defined as superhydrophobic when its θ_R is >90° and $\theta_A \geq 145°$. In this definition, a superhydrophobic surface will have no affinity with water. It will never be sticky as its θ_R has to be >90°.

This methodology has been extended to define oleophilicity/oleophobicity for hexadecane. A surface can be defined as oleophilic when its hexadecane θ_R is <124° and oleophobic when the hexadecane θ_R is >124°. The finding clearly demonstrates the incorrect presumption that the 90° cutoff is universal for all liquids. Essentially, the philicity/phobicity cutoff is dependent of the surface tension of the liquid, the lower the surface tension, the larger the θ_R cutoff.

8.3 Outlook and Recommendations

8.3.1 Surface Characterization

One key weakness in surface science is the correct interpretation of contact angle data. This is partly due to the absence of any standardized protocol where reliable, high-quality data can be produced and partly due to our insufficient understanding of the interactions between liquids and solid surfaces. By correlating the wetting force and adhesion force with contact angle data, the connections between wettability and adhesion with the advancing and receding contact angle are established. Based on these correlations, we recommend that surface should be characterized by its wettability, adhesion, and stickiness. Wettability is measured by the advancing contact angle. Adhesion is governed by the receding contact angle. As for surface stickiness, it can be predicted from the sliding angle; the larger the sliding angle, the stickier the surface [26]. It is important to point out that this recommendation gets its origin from the movement of the contact line during liquid–solid interaction. For wettability and adhesion, there is only one moving contact line. They are the advancing and receding contact line, respectively. An interface is formed when wetting occurs and that interface is eliminated when the liquid is fully receded. In the case of lateral mobility or sliding, both advancing and receding contact lines are moving simultaneously, and the wettability (or contact area) of the drop remains unchanged.

Contrast to the common belief, both static contact angle and CAH are not in the recommendation list. We however feel that static contact angle will still be the most measured contact angle in the future owing to its simple measurement procedure. It can be used as a surrogate for the advancing contact angle to predict wettability as the two angles usually track very well. Very likely it will be used as a screening or quality-control tool in the surface development labs. Advancing, receding, and sliding angles will play a larger role in the future for in-depth surface characterization and wetting study.

As for CAH, it is neither predicting wettability nor adhesion and stickiness. Although the mechanism for CAH is better understood, significant work remains particularly its role in surface characterization and wetting.

8.3.2   Guidelines for Best Practices

Many pioneering surface researchers were aware of the importance of surface cleanliness and liquid purification to the quality of the contact angle data they were generating. Before the design of the first goniometer in 1960, many different apparatuses were used to measure the different contact angles. While instrumentation has continued to improve, the desire to improve the consistency and reliability of contact angle measurement continues. About 40 years ago, Good [30] called for the establishment of standardized protocols for contact angle measurements, and the call was not answered. In the following, a set of general guidelines are provided. We hope that this will stimulate a more serious conversation in the research community as well as the authority or leader(s) to answer Good's calling.

Static contact angle. We recommend the use of a 5-μL droplet for general routine testing. Goniometer should be placed on a vibration-free table in a temperature/humidity-controlled lab. The deposit of the liquid droplet should be as gentle as possible and allows time for the drop to reach the static state before drop profile is captured and analyzed. If necessary, video can be used to record the wetting process and confirm the waiting time. The entire measurement procedure should be repeated 5–10 times in fresh area each time to ensure procedural consistency and sample homogeneity.

If a volatile liquid is used, the measurement should be carried out in a close chamber. Smaller drop size may be required when testing with low surface tension, high-density liquids, or on super repellent surfaces, where drop shape distortion by gravity is known to occur.

Advancing/receding contact angle. Advancing and receding contact angle should be determined with the drop expansion/contraction method. Typically, the measurement starts with a small sessile droplet (~2 μL), and small amount of test liquid is added at a very slow rate, e.g., ~0.2 μL/s. The drop profile is captured as the drop is expanded up to ~20 µL. The flow of the test liquid is then reversed, and small amount of liquid is withdrawn through the needle at the same, slow rate (~0.2 µL/s). The contact angle during receding is captured and analyzed.

The use of the titling plate method to determine advancing and receding contact angle is not recommended. The technique is known to have a limited range and measurement with large hysteresis sample is erratic [31, 32].

Sliding angle. Sliding angle is determined by first placing a sessile droplet on a horizontal surface followed by tilting the surface very slowly (~1°µL/s) till the drop starts to slide. The driving force for drop sliding is gravity, and the measured angle is strongly dependent on the drop mass. Based on available literature data, we recommend sliding angle to be performed with drop volume ranging between 5 and 10 µL. This range appears to be the most sensitive region to acquire drop mobility information. It would be a misleading conclusion if one claims a surface sticky when a small drop, e.g., 1 µL, is used to determine the sliding angle. The same would be true if one claims a surface slippery when a very large drop is used to determine the sliding angle.

Documentation. We highly recommend authors to provide the following information when publishing their contact angle results in the literature: apparatus, drop volume, dispense procedure, ambient condition, and drop profile curve fitting software. For super repellent surfaces, sticky droplets or super slippery surfaces, it does not hurt to report the sessile drop images or supply videos in the supplemental information section. Transparency would provide a trustworthy environment, which should facilitate data exchange, stimulating communications and collaboration.

References

1. Young T (1805) An essay on the cohesion of fluids. Philos Trans R Soc Lond 95:65–87
2. Zisman WA (1964) Relation of the equilibrium contact angle to liquid and solid constitution. In: Fowkes F (ed) Contact angle, wettability, and adhesion, advances in chemistry. American Chemical Society, Washington, DC, pp 1–51
3. Dupre A (1869) Theorie Mechanique de la Chaleur. Gauthier-Villars, Pairs, p 369
4. Gibbs JW (1928) Trans Connecticut Acad Arts Sci 1876–1878, 3; "Collected Works", vol. 1. Longmans, Green, New York
5. Rayleigh L (1890) On the tension of water surfaces, clean and contaminated, investigated by the method of ripples. Lond Edinb Dublin Philos Mag J Sci 30:386–400
6. Pease DC (1945) The significance of the contact angle in relation to the solid surface. J Phys Chem 49:107–110
7. Macdougall G, Ockrent C (1942) Surface energy relations in liquid/solid systems. 1. The adhesion of liquids to solids and a new method of determining the surface tension of liquids. Proc R Soc Lond A 180:151–173
8. Bartell FE, Hatch GB (1934) Wetting characteristics of galena. J Phys Chem 39:11–24
9. Bartell FE, Wooley AD (1933) Solid-liquid-air contact angles and their dependence upon the surface condition of the solid. J Am Chem Soc 55:3518–3527
10. Johnson RE (1959) Conflicts between Gibbsian thermodynamics and recent treatments of interfacial energies in solid-liquid-vapor systems. J Phys Chem 63:1655–1658

11. Cwikel D, Zhao Q, Liu C, Su X, Marmur A (2010) Comparing contact angle measurements and surface tension assessments of solid surfaces. Langmuir 26:15289–15294

12. Meiron TS, Marmur A, Saguy IS (2004) Contact angle measurement on rough surfaces. J Colloid Interface Sci 274:637–644

13. Della Volpe C, Maniglio D, Morra M, Siboni S (2002) The determination of a 'stable-equilibrium' contact angle on a heterogeneous and rough surfaces. Colloid Surf A Physicochem Eng Asp 206:47–67

14. Mettu S, Chaudhury MK (2010) Stochastic relaxation of the contact line of a water droplet on a solid substrate subjected to white noise vibration: role of hysteresis. Langmuir 26:8131–8140

15. Joanny JF, de Gennes PG (1984) A model for contact angle hysteresis. J Chem Phys 81:552–561

16. Johnson RE, Dettre RH (1964) Contact angle hysteresis 1. Study of an idealized rough surfaces. In: Fowkes F (ed) Contact angle, wettability, and adhesion, advances in chemistry. American Chemical Society, Washington, DC, pp 112–135

17. Neumann AW, Good RJ (1972) Thermodynamic of contact angles 1. Heterogeneous solid surfaces. J Colloid Interface Sci 38:341–358

18. Chen YL, Helm CA, Israelachville JN (1991) Molecular mechanisms associated with adhesion and contact angle hysteresis of monolayer surfaces. J Phys Chem 95:10736–10747

19. Extrand CW, Kumagai Y (1997) An experimental study of contact angle hysteresis. J Colloid Interface Sci 191:378–383

20. Wenzel RN (1936) Resistance of solid surfaces to wetting by water. Ind Eng Chem 28:988–994

21. Cassie ABD, Baxter S (1944) Wettability of porous surfaces. Trans Faraday Soc 40:546–551

22. Forsberg PSH, Priest C, Brinkmann M, Sedev R, Ralston J (2010) Contact line pinning on microstructured surfaces for liquids in the Wenzel state. Langmuir 26:860–865

23. Kanungo M, Mettu S, Law KY, Daniel S (2014) Effect of roughness geometry on wetting and dewetting of rough PDMS surfaces. Langmuir 30:7358–7368

24. Choi W, Tuteja A, Mabry JM, Cohen RE, McKinley GH (2009) A modified Cassie-Baxter relationship to explain contact angle hysteresis and anisotropy on non-wetting surfaces. J Colloid Interface Sci 339:208–216

25. Nosonovsky M (2007) On the range of applicability of the Wenzel and Cassie equations. Langmuir 23:9919–9920

26. Samuel B, Zhao H, Law KY (2011) Study of wetting and adhesion interactions between water and various polymer and superhydrophobic surfaces. J Phys Chem C 115:14852–14861

27. Roach P, Shirtcliffe NJ, Newton MI (2008) Progress in superhydrophobic surface development. Soft Matter 4:224–240

28. Law KY (2014) Definitions for hydrophilicity, hydrophobicity, and superhydrophobicity: getting the basic right. J Phys Chem Lett 5:686–688

29. Law KY (2015) Water interactions and definitions for hydrophilicity, hydrophobicity, and superhydrophobicity. Pure Appl Chem 87(8):759–765

30. Good RJ (1977) Surface free energy of solids and liquids. Thermodynamics, molecular forces, and structures. J Colloid Interface Sci 59:398–419

31. Krasovitski B, Marmur A (2005) Drops down the hill. Theoretical study of limiting contact angles and the hysteresis range on a tilted plate. Langmuir 21:3881–3885

32. Pierce E, Carmona FJ, Amirfazli A (2008) Understanding of sliding and contact angle results in tilted plate experiments. Colloids Surf A Physicochem Eng Asp 323:73–82

Index

© Springer International Publishing Switzerland 2016

K.-Y. Law, H. Zhao, *Surface Wetting*, DOI 10.1007/978-3-319-25214-8